T0075430

Addressing the Climate in Modern Age's Construction History

Carlo Manfredi
Editor

Addressing the Climate in Modern Age's Construction History

Between Architecture and Building Services Engineering

Editor
Carlo Manfredi
Politecnico di Milano
Milan, Italy

ISBN 978-3-030-04464-0 ISBN 978-3-030-04465-7 (eBook)
https://doi.org/10.1007/978-3-030-04465-7

Library of Congress Control Number: 2018965442

This Springer imprint is published by the registered company Springer Nature Switzerland AG
The registered company address is: Gewerbestrasse 11, 6330 Cham, Switzerland

Preface

This book aims to cast light on the environmental control in buildings, since the seventeenth century onwards. Even before building services became the main hallmark in buildings, in order to face sanitary and comfort increasing needs, pioneering experiences had improved the design skills of professionals. After being for ages decided, most of all, by construction passive features, indoor climate went under the setting of installations and plants, representing the most significant shift of paradigm in the modern age's construction history.

This change being not without consequences, the essays collected are focused to show how deep is the connection between architectural design, comfort requirements and environmental awareness throughout the long nineteenth century.

Taken into account the differences between different European Countries, the focused topic will be of interest to architects and designers which are concerned with environmental design, in order to define a deeper knowledge of heritage behaviour face to the climate requests, particularly going towards a future in which energy savings and fuel consumption reduction will border our ways.

The volume is composed of nine contributions, written by 11 authors:

- Chapter "Indoor Climate, Technological Tools and Design Awareness: An Introduction" by book editor Carlo Manfredi, which also serves as Introduction of the volume;
- Chapter "Heating Verona in the Nineteenth Century. From the Fireplace to the Hot Water Systems" by Marco Cofani and Lino Vittorio Bozzetto;
- Chapter "The Alte Pinakothek" by Melanie Eibl-Bauernfeind;
- Chapter "Two Early Examples of Central Heating Systems in France During the 19th Century" by Emmanuelle Gallo;
- Chapter "Camillo Boito and the School Buildings Indoor Climate in the Unified Italy (1870–1890)" by Alberto Grimoldi and Angelo Giuseppe Landi;
- Chapter "Tradition and Science: The Evolution of Environmental Architecture in Britain from 16th to 19th Century" by Dean Hawkes;
- Chapter "Not Just a Summer Temple: The Development of Conservation and Indoor Climate in Nationalmuseum, Sweden" by Mattias Legnér;

- Chapter "Asserting Adequacy: The Crescendo of Voices to Determine Daylight Provision for the Modern World" by Oriel Prizeman;
- Chapter "The Houses of Parliament and Reid's Inquiries into User Perception" by Henrik Schoenefeldt.

Milan, Italy

Carlo Manfredi

Contents

Indoor Climate, Technological Tools and Design Awareness: An Introduction

Carlo Manfredi

Abstract This contribution explores the mutations and adaptations provided by the appearance of building services engineering on the architectural scene, since the end of eighteenth century. It try to open new perspectives on construction history by depicting the state-of-the-art information of recent studies and researches in the field of environmental approach to the architectural history, particularly in health and comfort technologies. It shows how mechanisms that connect (ancient and new) buildings and the environment should be well-mastered, and steer our behaviour in future choices, especially rethinking new uses of built heritage. As well as in structural interventions, new arrangements shouldn't obliterate the past lives, material tracks and marks on the architectural legacy.

Undeniably, by now, history is the history of material culture. History of the objects: built, produced, and then wanted of refused, preserved, lost or discarded. Moreover: since production is always more abundant than necessary, there is always more of what remains and must be discarded, than of what is preserved. At the end, "History is junk! A society may be characterized by the contents of its drains, waste piles, water-closets, graveyards or chimneys. Smoke and soot have generally been considered too tenuous for archaeological examination, and the historian has usually held an orange before his nose and made but fleeting reference to fleas, fumes, sweat and dung" (Brimblecombe 1987[1]).

Actually, especially in recent years, attempts have been made to keep the orange at a distance. Some efforts by historians have focused on the less noble productions,

[1]P. 1.

This contribution was translated by Pierfrancesco Sacerdoti.

C. Manfredi (✉)
Politecnico di Milano, Milan, Italy
e-mail: carlo.manfredi@polimi.it

C. Manfredi (ed.), *Addressing the Climate in Modern Age's Construction History*,
https://doi.org/10.1007/978-3-030-04465-7_1

whose impact modified and modifies the environment which hosts man's life, and which modifies it to such an extent that it not possible not to notice.[2]

At least since the experience of the *Annales*, with methods taken from the "hard sciences", especially from archaeology, experimental research regarding architecture and construction has been carried out on the traces, even the most tenuous, which recount the evolution of material production, so full of information on the culture which produced it.

History of architecture has partly been an exception, leading to single out in fields which are perhaps tangential (the history of construction, restoration and the history of restoration) the context in which to investigate the lines of research which make use of the sources most related to a contemporary and not evenemential conception of history.[3] It is on the basis of these contiguous but only partially superimposable fields that archaeology, and the historic methods most adherent to today's *forma mentis*, have been capable of facing issues related to the changes of use and living, which are the issues which have affected construction in modern times the most. Naturally, the forms of living and their evolution have been analysed by historians and make up a source of information useful for the understanding of social and behavioural mutations. But they have been analysed in a historical perspective and with methods tied to social history rather than to architectural history. I am thinking about the historiographical tradition which has evolved on the traces of an evolution of living, which has supplied noteworthy sparks for social history, and which has defined the times and spaces of a relatively recent process, codified and acknowledged since the eighteenth century.

In-depth research work, well known by now, has combined high cultural levels with widespread dissemination, as can be seen in Georges Duby's *Histoire de la vie privée* and Michelle Perrot's recent *Histoire de chambres*. During the eighteenth century, alongside and perhaps before the acquisitions due to the Industrial Revolution, a series of mechanisms and customs which led to a specialisation of domestic space and changed the balances within the private house came about (Eleb-Vidal and Debarre-Blanchard 1989; Perrot 1989). This revolution was welcomed in the next century also for public spaces.

Even if they do not exhaust the category of the *dispositifs* (apparatus) which monitor comfort, the mechanical systems for buildings are the symbol of this change.[4] It was a radical change. Since the end of the eighteenth century, modernity

---

[2]Although, according to Brimblecombe, those who "have handled the less polite aspects of the human society" are "irreverent few", p. 1; furthermore, see: Schott et al. (2005), Guillerme (2007) and Sori (2001).

[3]Indeed, this concept was focused on by the *Ecole* of the «Annales d'histoire économique et sociale», the magazine founded by Lucien Febvre and Marc Bloch in 1929.

[4]The French word *dispositif*, as used by Michel Foucault, has been translated with *apparatus*, as defined by the philosopher Giorgio Agamben: "[…] I shall call an apparatus literally anything that has in some way the capacity to capture, orient, determine, intercept, model, control, or secure the gestures, behaviors, opinions, or discourses of living beings. Not only, therefore, prisons, madhouses, the panopticon, schools, confession, factories, disciplines, judicial measures, and so forth

began to split up reality into elements, which had to fulfil a function or satisfy a need, thus loosening the relationship with the whole, giving specific tasks to the tools which could carry them out, disarticulating the overall meaning to entrust it to expert and specialised elements. The scientific disciplines specialised and moved apart increasingly, they delved into a specific aspect and neglected the overall vision.

This is evident for those who deal with construction. The separation between structure and enclosure is perhaps the most evident of the changes introduced, and certainly the most studied along with the distinction between "servant" and "served" spaces. The potentials offered by the new materials—iron, steel, glass, and later concrete—fulfilled the increasingly insistent need to separate the functions of living—and, before that, to produce materials and construct buildings—from the general conception of construction. The category of mechanical systems fulfilled the specific task of improving comfort. First came the protection of decency, so to speak, related to the behaviours which made some body functions appear intolerable from a certain point onward: they were confined to specific areas and delegated to designated devices (Vigarello 1985; Wright 2005). Then came the strategies for the reduction of toil and discomfort, especially aimed at the improvement of environmental comfort. Hygienic needs which had effects also on the urban environment, and which entailed the construction of distribution networks for drinking and waste water, connected to the terminals in private homes, were developed. And then came the centralisation of heating systems, in which the annoyance of the management of the hearth was assigned to a room and specific worker, the stove setter. Later still, transportation networks were introduced: for energy and people, to feed a network of public and private buildings, suitable for the fulfilment of the functions of a complex urban community.

It seems to me that investigation on how these objects changed in time, on how they fed the collective imagination, and were then largely abandoned after a short and often inglorious trial run, and on how they were then interpreted differently with respect not so much to external influence but mostly to the cultural needs they expressed (or which were attributed to them), may be a significant step for the knowledge of recent history, and lastly, for the elaboration of today's mechanisms which are probably not fully understood.

The idea for this book sprang from these initial reflections. An attempt has been made to collect some studies related to the line of research which has become known as 'Construction History', especially those regarding comfort in buildings, such as temperature and humidity conditions, ventilation, natural or artificial lighting, sanitary facilities, and to relate them with the changes of the conditions

---

(whose connection with power is in a certain sense evident), but also the pen, writing, literature, philosophy, agriculture, cigarettes, navigation, computers, cellular telephones and–why not–language itself, which is perhaps the most ancient of apparatuses–one in which thousands and thousands of years ago a primate inadvertently let himself be captured, probably without realizing the consequences that he was about to face" (Agamben 2009, p. 14).

which determined the construction of buildings at a given moment (which in this case is the nineteenth century).

Within this reflection there are a series of issues which have been investigated until now from points of view almost always different from that we wish would emerge here, and which offer, however, ideas which appear to herald new developments, especially when one thinks of the reciprocal interrelations.

It is worth retracing some of the events which have concerned European capital cities since the nineteenth century. With the enormous urban migration due to the exponential growth of production, London was the first city to experience significant changes in every sector. The size of the settlement led to "changes in London's drainage system which occurred in the early nineteenth century and which represented a significant change from the arrangements which had prevailed at least since medieval times" (Halliday 1999[5]). Indeed, London managed the waste water of buildings with cesspools until 1815: "[…] it was penal to discharge sewage or other offensive matter into the sewers, which were intended for surface drainage only. The sewage of the Metropolis was collected into cesspools which were emptied from time to time and their contents conveyed into the country for application to the land" (Halliday 1999[6]). Later, due also to the fact that the country was moving farther and farther away from the city centre (because of urban growth) and to the consequent and imaginable increasingly greater difficulty of disposal, emptying of the sludge in the network which had been used until then only for surface drainage was authorized. Among the other factors which contributed to bring about the situation to come to a head, and to the founding of a unified Metropolitan Commission of Sewers in 1848, "[…] the first was the growing popularity, from the late eighteenth century, of the water-closet" (Halliday 1999[7]). The new apparatus required a much larger amount of running water than that used previously, and thus the cesspools became useless and were abandoned. All the sewage network flowed into the secondary rivers which cross London, and which flow into the Thames. This caused deleterious effects to an imaginable extent: WCs' "widespread adoption had devastating consequences for the antiquated sewers to which they were connected" (Halliday 1999[8]). Moreover, it is useful to recall that the Thames was the greatest source of drinking water back then. The consequences were not long in coming: London suffered four pandemics of cholera between 1831 and 1866. In his *Report on the Sanitary Condition of the Labouring Population of Great Britain*, Edwin Chadwick, promoter of the Sanitary Movement, focused on four main issues: "The first was the relationship between insanitary conditions and disease. The second was the economic effects of poor living conditions as manifested in the creation of 'cholera widows', orphans and those rendered by disease incapable of work, all of whom had to be supported by the Poor rates. The third

[5]P. 26.

[6]*The Builder*, August 16th 1884, p. 215; cited at p. 29.

[7]P. 42.

[8]P. 49.

theme concerned the social effects of poor living conditions—intemperance and immorality as well as disease", the fourth one being "the need for new systems of administration to bring about a reform" (Halliday 1999[9]).

The situation became increasingly critical and exploded in the summer of 1858, the season of the Great Stink. With temperatures above average, in a city whose size was growing exponentially, the conditions became unbearable. The press recalls how prime minister Disraeli, after having suspended the meeting of the committee he belonged to, left shortly afterwards with the other members of the committee, "with a mass of papers in one hand, and with his pocket handkerchief applied to his nose" because of the unbearable stench (State and Church 1858[10]). And this even though the curtains on the river side of the building were soaked in lime chloride to overcome the smell.

Following imaginable social and political tensions, the problems were solved by building a network for the disposal of sewage, beginning from the Crossness Pumping Station, designed by Joseph Bazalgette and Charles Henry Driver, which opened in 1865. The complex operation modified the relationship between city and river, also in relation to the construction of the embankments on the Thames, which contain sewers all the way to the collecting stations.

Peter Brimblecombe offers us a parallel, and partly superimposed, reading of the events related to the emissions and exhausts due to combustion; he stigmatises the stink as "popular perception of the origin of diseases" (Brimblecombe 1987[11]) since the middle ages, following the path which leads to the pre-scientific acceptance of the miasma theory.

The polluting factors in London had intensified at least since the seventeenth century, not only because of the growth of industrial and handcrafted production, which remained limited until the next century, but also in relation to the increase of domestic heating due to the acute phase of the Little Ice Age approaching its climax, and because of the gradual passage from wood to charcoal and then to coal, both for houses and factories.

Today we know that the pollution caused by the combustion of charcoal is due to the production of sulphuric acid. A debate, which developed in England in the mid-seventeenth century, evaluated the positive and negative elements of London's climatic and environmental conditions: Kenelmy Digby "was particularly worried about the damage these volatile and corrosive salts caused to the lungs, claiming that their presence in the London air was responsible for the high incidence of deaths from pulmonary complaints in the city. It was argued that more than half those who died in the metropolis did so from 'ptisical and pulmonary distempers, spitting blood from their ulcerated lungs'. The air of London was worse than that of Paris or Liège, so people with weak lungs and plenty of money were advised to live on the continent" (Brimblecombe 1987[12]). A different opinion was expressed by

[9]P. 38.

[10]Pp. 423–424.

[11]P. 9.

[12]P. 46.

"Sir William Petty, FRS (1623-87) who thought the air of London was more wholesome, and that there were fewer deaths than in Paris, because the fuel of London was cheaper and not as bulky, being a wholesome sulphurous bitumen. No doubt he thought that a fuel that was cheap and compact would be more readily purchased and lead to warm, safe interiors in the winter. Digby would have disagreed, although he was willing to acknowledge that Paris suffered excessively from stinking dirt which mingled with the air. He insisted that the Parisian air was not as pernicious as that of London. Most of the air pollution in Paris arose from the rotting wastes in the sewers of the city but, as previously noted, such olfactory pollutants usually received far less attention than those from coal burning" (Brimblecombe 1987[13]).

Actually, the situation in Paris was probably neither better nor worse. Judging from Emile Beres's description, also here the conditions of the working class must have been extremely critical: "Employers are often guilty of unpardonable carelessness with respect to the employed… Here a deleterious atmosphere, which ought to be carefully purified, is imprudently allowed to be inhaled; there a poison, which ought to be handled with precaution, is allowed to penetrate every pore…" (Billington and Roberts 1982[14]).

It was the productive structure of a capital city that made the precarious environmental conditions necessary. Most urban industrial activities "impliquent les mêmes conditions techniques: une forte humidité, donc une faible célérité de l'eau et de l'air; et une longue macération des produits de base, une putréfaction, conditions interdépendantes, qui exigent la stagnation des eaux. L'usage de l'eau, base des activités urbaines, se transforme sensiblement: ce n'est plus tant sa dynamique, déjà totalement mise en valeur par les moulins, que sa stagnation, sa nébulosité, qui est exploitée par la ville de l'Ancien Régime" (Guillerme 1983[15]). Humidity favoured the processes of alteration and oxidation for the production of saltpetre, vital for the war industry, but also for leather and paper production, and many others. "La putréfaction adhère à la ville comme l'humidité qui colle à cette longue période de refroidissement climatique. L'une et l'autre ont déterminé la richesse de la ville; plus une ville est puante, plus elle est riche. La putréfaction est un support de l'urbanisation, comme le sacré au Bas-Empire" (Guillerme 1983[16]). The construction of a sewer system which flanks the distribution networks would later be one of the decisive elements for the success of Haussmann's reform (Loyer 1987[17]).

John Evelyn, in exile with Digby in France during Oliver Cromwell's protectorate, known for his *Fumifugium or The Inconveniencie of the Aer and the Smoak of London Dissipated* (1661), "advanced ideas on the need for extensive urban

[13]P. 46.

[14]P. 184.

[15]P. 150.

[16]P. 179.

[17]P. 187 and ff.

planning to make cities attractive places for their citizens" (Brimblecombe 1987[18]). In *A Parallel of Architecture* (1664), he "found the streets of London composed of a congestion of misshapen and extravagant houses in streets much too narrow" (Brimblecombe 1987[19]). He admitted that firewood and charcoal polluted less than fossil coal, and he realised that many emissions in the city were caused by the activity of small craft businesses, such as brewers, lime-burners, blacksmiths, "salt and sope-boylers". The various negative effects caused by smoke are described in the *Fumifugium*: "churches and palaces looked old, clothes and furnishings were fouled, paintings were yellowed, the rains, dews and water were corrupted, plants and bees were killed, and human health and well-being were ruined" (Brimblecombe 1987[20]). There was certainly an increase in the mortality rate, and a "General decline in health" (Brimblecombe 1987[21]). But since there is 'no smoke without fire' and 'no fire without smoke', Evelyn proposed moving the industries out of the city, in an area east of London, probably Shooter's Hill, which could easily allow the transport of the workers with river connections. To conclude, "odiferous flowers, plants and hedges were to be planted about the city" (Brimblecombe 1987[22]).

On the strength of his proposal, Evelyn was designated to prepare the draft of a bill to improve the quality of the air, which however was never discussed in parliament, and which thus was probably abandoned rapidly. Also following the great opportunity of rebuilding a healthier city, after the Great Fire of 1666, Evelyn's projects had no significant consequences.

In the same years, John Graunt, a draper, wrote up a detailed analysis on the mortality rate: *Natural and political observations... Made upon the Bills of Mortality in 1662*, which "remains a pioneering work on demography" (Brimblecombe 1987[23]). The bills of mortality were drafted weekly from charts which were written in each parish by "ancient Matrons", who accurately examined each corpse trying to describe the causes of death. Despite the dishomogeneous preparation, and the often less than basic medical skills, it was possible to establish series of credible data, useful to identify some of the causes due to environmental conditions. For example, the increase of rickets (which could become a cause of death), due to malnutrition (lack of vitamin D) or to insufficient exposure to sunlight.

The grey London skies became from this moment on a literary subject: "thick skies and cloudy weather" (Brimblecombe 1987[24]). Although it is difficult to relate an effect (also uncertain, due to the detection method) directly to its causes, which

[18]P. 47.

[19]P. 47.

[20]P. 49.

[21]P. 49.

[22]P. 50.

[23]P. 52.

[24]P. 54. H. R. Bentham visited London in the second half of the seventeenth century.

may be multiple, from the malnutrition due to the increase of prices, to the scarse sunlight, which could be caused both by air pollution and by the "almost complete covering of the skin required by the cold weather of the little ice age" (Brimblecombe 1987[25]). In any case, it is certain that the mortality rate was much higher in London than in the rest of the country.

Due to all these implications, "[...] the air pollution is not simply affecting our health, it affects our entire wellbeing and moral judgement" (Brimblecombe 2010[26]). The "smoke doctors", among which were Prince Rupert, Dalesme, Gauger, and Rumford, went wild searching for a technical solution which would allow a clean combustion and fuel savings (Wright 1964[27]). As will be seen later on, the quality of air (and the attention dedicated to it) was not considered simply a scientific and measurable fact: "De même que l'ensemble des innovations technologiques de la période précédente calquait fidèlement l'imaginaire social, l'approche des scientifiques (des «philosophes») puis des techniciens reflètera le dégoût de la mort, de la décomposition de la matière, et de la putréfaction qui s'empare de l'Occident à la fin du XVIIIe siècle. A la «répugnance à représenter ou à imaginer la mort et son cadavre» comme l'écrit Ariès, fait pendant une nouvelle technologie urbaine fondée sur la dynamisation de l'eau, l'accélération des transformation chimiques et plus généralement le développement de la chimie de synthèse" (Guillerme 1983[28]). In spite of the pionnering efforts of personalities such as Benjamin Franklin and Count Rumford, the growing awareness which tried to oppose pollution which was totally avoidable, obtained concrete results only with the popular movement which led to the approval of the Smoke Nuisance Abatement Act in 1853 (Brimblecombe 1987[29]). In the parliamentary debate the cause was supported by Lord Palmerston, who later became prime minister.

Despite the agreement between the social conscience and the political lobby, the efforts later undertaken by the manufacturers for the containment of the emissions were carried out only to reduce consumption and to yield immediate cost savings. It was not the fear of the effects of pollution on human health which determined the gradual refinement of the techniques of indoor climate control.

On the construction scale (that of buildings), there is literature which refers to the technical developments which have made indoor climate control possible in a modern sense, that is through the application of a series of technological tools which have evolved step by step, and which have allowed the fulfilment of a specific need, the implementation of a given function. It is a non-linear process, which developed fully during the nineteenth century. The production of heat and the ways of conveying it became the object of in-depth research. From Lavoisier's

[25]P. 54.

[26]P. 228.

[27]P. 83 and ff.

[28]P. 188; taken from P. Ariès, *Essai sur l'histoire de la mort en Occident du Moyen Age à nos jours*, Paris, 1975.

[29]P. 108.

intuitions, which established the equivalence between heat and energy, to Rumford's studies, to "Joule's first direct determination of the mechanical equivalent of heat" (Billington and Roberts 1982[30]) in 1843, passing through Tredgold's treatises on steam and Péclet's (imprecise) measurements, many people dealt with heat loss and heat transfer. Despite Joseph Fourier fine-tuning of equations to describe the ways of conveying heat as early as the 1820s, the projects carried out had an experimental character for a long time yet. A propos of one of the most successful central heating systems, also from a commercial standpoint, the famous High Pressure Hot Water system patented by Angier March Perkins in 1840, we know that, despite the many descriptions made by Richardson (1837[31]) there is "no clue as to design procedure" (Billington and Roberts 1982[32]). And, besides, still in 1904 "Perkins' high pressure systems were also designed on an empirical basis" (Billington and Roberts 1982[33]).

Despite of the dissemination of centralised heating systems, climate control, especially in private homes, and more consistently in the lower classes, was still entrusted to the fireplace. A detailed analysis of the technical progress which allowed the evolution of technology is dealt with in texts which differ but have common aims: Billington and Roberts, with a meticoulous overview of the so-called modern age inventions, and Elliot with a series of applications developed in the field of construction (Billington and Roberts 1982; Elliott 1992).

A series of buildings, which expressed the success of the bourgeois society over the course of the nineteenth century, has also been analysed separately. These are the objects which produced change. The enquiry, which highlights how the evolution of the ways of production accompanied the change of the functions, was carried out more easily in the large capitals, such as Paris, London and Berlin. Alongside the change made in the networks (transport, transport of energy, distribution of drinking water, disposal of waste water), also the buildings underwent a change which went hand in hand with the emerging of new functions in collective buildings, earlier than in the increasingly complex organisation of apartment buildings (Guillerme 1983, 2007). After Giedion's and Banham's pioneering studies, the attention of scholars has begun to focus on the mechanical systems of historic buildings only since the 1970s, with a series of articles published unevenly by scholars belonging to the Anglo-Saxon context. Robert Bruegmann's article *Central Heating and forced ventilation* compares the evolution of air, steam and water heating systems over the course of the nineteenth century, starting from the scientific literature of that period (Bruegmann 1978). This article has the merit of acknowledging the close relationship between heating and ventilation.

The series of articles published in the "Architectural Journal" and collected by Dan Cruickshank with the title *Timeless Architecture*, is perhaps the first example

[30]P. 485.

[31]Perkins's systems were widely disseminated beginning from the 1830s, not only in Great Britain.

[32]P. 118.

[33]P. 489.

of a new systematic approach, which remained in an embryo state, however, still for a long time. Five buildings are analysed in Cruickshank's book: they are apparently very different from the point of view of form and function, but share a recognisable quest for comfort in modern terms. A parish church, a very exclusive private club, a natural history museum, a gallery for paintings and a private residence, all built between the nineteenth and the beginning of the twentieth century (Cruickshank 1985).

In St. Andrew's Church in Roker, designed by Edward Prior at the beginning of the twentieth century, the architect "not only wanted the forms and materials of his churches to express their spiritual purpose, he also wanted them to be comfortable. To achieve this, he incorporated a comprehensive warm-air heating system. This consisted of an extensive hypocaust arrangement running in horizontal shafts beneath the floor of the nave" so that "warm air was delivered through floor grilles and outlets in the side walls in the space between the top of the panelling and the window sills" (Hawkes 1985[34]). From the beginning of the nineteenth century, many religious buildings were equipped with similar systems, starting from the adaptation of the older churches (Campbell et al. 2012).

Charles Barry, the architect known for the reconstruction of the Houses of Parliament in London after the 1834 fire, was always very attentive to the implications of the indoor environment in planning buildings. In the project for the Reform Club (1839) in London he took the model of the Italian Renaissance palace (Palazzo Farnese) and adapted it "to suit the English personality and climate" (Olley 1985[35]). This "unmistakably English" building was deliberately "reticent and aloof", and combined the most refined classicism with the cutting-edge technology of the time. The lighting system, conceived as a gas system from the start (a gas distribution network was present in the Pall Mall area since the 1820s), contributed to the ventilation of the rooms through aspiration, and was thus designed to integrate with the heating system. "The saloon sun burner, which could be lowered or raised with a system of counterweights located above the roof, had a shaft above which led up to a ventilator. The heat from the burner effectively ventilated the whole room by means of convection" (Olley 1985[36]). Barry developed this experimentation in the long construction process of the Houses of Parliament, in later years (see Henrik Schoenefeldt's contribution in this volume). The heating system of the Reform Club "consisted of a 5-horsepower engine which pumped water for household purposes, raised coals to the several apartments on the upper floors and drove the fan ventilator. The steam was condensed in cast iron heat exchangers where it heated the air [which] was driven [...] and let off into a series of distinct flues, governed by dialled valves or registers whereby it is conducted in regulated quantities to the several apartments" (Olley 1985[37]). The "background

[34]P. 18.

[35]P. 26.

[36]P. 36.

[37]P. 38.

heating" thus supplied had to be supported by further devices, such as fireplaces "which created both a focus and a graded thermal environment within the room" (Olley 1985[38]). A cooling system was introduced during redecoration work in 1878.

When, at the end of the 1871 World Fair, the area of South Kensington became available, new buildings for collective use were designed and built. One of them was the Natural History Museum, designed by Alfred Waterhouse, which opened to the public in 1881. It is a large building, whose mediaeval revival forms, made up of terracotta elements, clad a steel and Portland cement structure. Waterhouse made many drawings for the heating and ventilation system: "Indeed, the important picturesque elements of the design were needed for the system of ventilation. The two towers to the rear were to act as smoke flues for boilers and to carry away vitiated air. The corner turrets of the end pavilions and south towers were to be outlets for the ventilation of the three floors [of] galleries to the south" (Olley and Wilson 1985[39]). As in the case of the dispute concerning the House of Commons, also on this occasion "any suggestion from the Office of Treasury for the removal or the reduction of height of any of the towers was met with arguments concerning the consequent impaired efficiency of ventilation" (Olley and Wilson 1985[40]).

The system chosen by Waterhouse was simple and well tested: "Three steam boilers were sunk into the ground in the basement beneath […]" the Museum, and they "supplied steam chests, which transferred their heat to hot water that was then piped around the building in channels beneath the basement floor" (Olley and Wilson 1985[41]). Natural light, believed to be more adapted for the exhibition needs, was also privileged (Cook and Hinchcliffe 1995, 1996).

All the building designed and built in South Kensington around 1870, like the Natural History Museum, were supplied with adequate mechanical systems. But experimentation had begun more than half a century earlier.

When it was finished, in 1813, the Dulwich Picture Gallery was the first building in Great Britain to have the single aim of collecting and exhibiting works of art.

The structure was finished in mid-August 1812, but the walls still had to be plastered and "the suspended floor was awaiting the installation of the steam heating system in its underground duct by Boulton and Watt" (Davies 1985[42]). As was often the case in these pioneering installations, during those years of "tremendous advancements in central heating methods" (Willmert 1993[43]), "this innovative system was to cause terrible problems in the first few years of the gallery's life, chiefly because it almost caused dry rot in timber floor" (Davies

[38]P. 38.

[39]P. 66.

[40]P. 66.

[41]P. 66.

[42]P. 85.

[43]P. 26.

1985[44]). Since the beginning, leaks due to the "sliding expansion joints" of the iron piping were detected, while the ongoing malfunctioning led to a survey in 1819, following which it was suggested "that the whole system should be taken out and sold" (Davies 1985[45]).

As in Barry's case later, John Soane made use "[...] of both traditional and modern heating methods. Fireplaces were an important element of his architecture because of their psychological connotations" (Willmert 1993[46]). The hearth preserves atavic reminiscences, tied to the meaning sedimented in the collective imagination (Bachelard 1938).

"While Soane embraced traditional heating methods, he also took advantage of the central heating systems emerging during his career. Steam, hot water, and hot air systems—heating innovation developed in Great Britain during the industrial revolution—provided new means of controlling the thermal environment, and Soane recognized and explored them for the comfort they afforded building occupants as well as for the design opportunities they presented" (Willmert 1993[47]). Indeed, the designs for the Bank of England Stock Office and the Dulwich Picture Gallery were important "experimental workshops" on the issue of comfort. Over the course of forty-five years, in the house-studio he lived in at Lincoln's Inn Field (and which, according to his will, became a museum after his death), Soane tried out three different centralised systems, and carried out many solutions tied to the thoughtful use of natural light.

The Hôtel Tassel (1892) combines the quest for comfort with the hygienic and sanitary needs which arose over the course of the century. As will be seen further on, the latter were certainly more decisive than the certainly present "healthy obsession" of its designer, Victor Horta. The result is noteworthy, thanks to the harmonious relationship between materials, structure and ornaments, and the meaning the building took on in the brief trajectory of the style known in French-speaking countries as *Art Nouveau*. A great deal of attention was dedicated to a new conception of living, in which the use of new building technologies had potentials that did not only regard form. Also, the mechanical systems were designed so they would be integrated in the building.

"The original heating system for the house consisted of a coal-fired furnace in the basement below the ground floor utility room" (Parrie and Dernie 1985[48]).

The balance of the building's organism was such that it became a recognised model. "Air for combustion and warm air distribution was drawn under the basement floor through a duct from the garden facade wall. Once heated, the air was ducted to the vestibule, the lightwells and the dining room" (Parrie and Dernie

---

[44]P. 85.

[45]P. 85.

[46]P. 26.

[47]P. 26.

[48]P. 101.

1985[49]). Also in this case, single heating devices were fundamental as integration of the centralised system: "Rooms on the upper floor were heated with gas and coal appliances. All rooms were ventilated by flues next to discharge appliance flues to increase the stack effect" (Parrie and Dernie 1985[50]; Loyer and Delhaye 1986).

Unlike the Hôtel Tassel, the contemporary, celebrated Glasgow School of Art, designed by Mackintosh at the turn of the century, was built with a rather traditional load-bearing wall structure. On the interior, however, the strict and innovative architectural language of the facades gave way to a gradual modulation of light, an "ever-changing lighting" suitable for the functions of the different spaces, and to a "principal environmental system […] absolutely of their days" (Hawkes 2008[51]). The attention payed to the lighting conditions of the studios for life drawing, which required a consistent supply of natural light from the north though large glass surfaces, made necessary a mixed system with hot air and cast-iron radiators, further integrated by fireplaces in the offices, to compensate the relevant heat losses. In the harsh Scottish climate "there was no inconsistency in incorporating domestic elements in an institutional building, if the nature and use made this appropriate" (Hawkes 2008[52]).

Well before the attempts to build central heating systems using water or steam, which became more and more reliable and efficient as the iron and steel technologies evolved, large hot air heating systems were installed in monumental buildings. In Saint Petersburg, the Russian capital founded by Peter the Great, the design of the city progressively developed over the course of the entire eighteenth century. At the death of the first Tsar who sought inspiration in the West, the government was ruled by czarinas Catherine I, Anna Ivanovna, Elisabeth I and especially Catherine the Great (from 1762 on). Among the many artists who worked for the court there was Bartolomeo Francesco Rastrelli, an architect of Italian origin born in Paris in 1700, "a brilliant man, an architect who mastered to the fullest the art and technology of the building" (Lerum 2016[53]). For the design of the heating systems in the court buildings, he put into practice the intuitions of Daniel Bernoulli, who was also often at court and was then empirically developing a theory on hydraulic flows: "Knowingly or not, in his [Rastrelli's] design for the great palace on the banks of the Neva, he applied Bernoulli's principles, integrating hundreds of channels inside the walls and a thousand chimneys on the roof" (Lerum 2016[54]).

At the beginning, such sophisticated skills were applied only to the size of the chimney flues and ventilation ducts of the rooms: "[…] each room with a fireplace

[49]P. 101.

[50]P. 101.

[51]P. 19.

[52]P. 23. Mackintosh showed a great sensitivity to indoor environmental issues also in private buildings: the Hill House in Helensburgh is described in Hawkes (2012, pp. 180–182).

[53]P. 26.

[54]P. 28.

was served by three channels built in the thick masonry walls and connected to the chimneys or hoods on the roof. While one channel served as a smoke chimney, the one next to it was used to replenish air consumed by combustion. The proximity of the fresh air channel to the smoke chimney caused the incoming air to be preheated. The third channel, with a grille near the ceiling, served as an exhaust duct for vitiated air" (Lerum 2016[55]).

This complex system was used alongside with the "Dutch stoves" which Peter the Great himself had seen in Europe, and which were imported from Holland and Meissen still in the first half of the nineteenth century. A centralised system was designed for the Winter Palace only after the devastating fire of 1837. The new system took advantage of the generous wall mass of the building, making channels over a metre and a half wide: "The system of channels inside thick walls is still the living backbone of the modern air conditioning system that serves the palace today" (Lerum 2016[56]). The *fortochka*, double-glazed windows, which permitted different degrees of regulation of the openings, and greater thermal isolation, contributed to the comfort of the palace.[57]

The system installed during the reconstruction of the building, which was completed within a year, was based on the use of the stoves invented by General Amosov a few years earlier, which consisted mainly of a heat-exchanger which could be placed in the basement. Eighty-four of these were installed. They fed a complex network of air channels (the existing ones were extensively reutilised) which served all the rooms on the upper floors. It was no longer necessary for servants to enter the reception rooms to bring wood or to clean and remove the ashes: "Instead, warm air magically heated and ventilated the palace quietly and invisibly" (Lerum 2016[58]).

Other buildings have been examined and rediscovered after the oblivion which followed their construction, caused by an attitude which was incapable of seeing the innovative aspects which were concealed behind their historicist forms. These buildings passed almost unnoticed, so to speak, because of their conformity to the rules of a type of architecture which, having to satisfy collective rather than individual needs, conformed to traditional models, often only apparently.

One of the most refined architects of the nineteenth century was Henri Labrouste, whose few realisations, sober and adequate to their context, were almost forgotten for a long time by the historiography of architecture. His limited notoriety for having built the Bibliothèque Sainte-Geneviève and refurbished the Bibliothèque Nationale, has been only partly related to the precocious mastery of the technical tools (regarding both the structures and the mechanical systems) which characterises his work. Indeed, "the mechanical systems and installations are among

[55]P. 28.

[56]P. 29.

[57]The distance between the two windows could span thirty centimetres. These windows are better known in technical literature with the German name of *Kastenfenster*, cfr. Huber (2012).

[58]P. 32.

the best-kept secrets of the Bibliothèque Sainte-Geneviève" (Lerum 2016[59]). Built to replace the obsolete structure of the Collège Sainte-Barbe, it was inaugurated in 1851. The generous openings which filtered the natural light were integrated by a gas lightings system which made it possible to keep the building open until ten o'clock in the evening. The heating was entrusted to a hot air system, whose ventilation was activated by the dozens of flames of the gas lamps. The hot air which accumulated in the higher part of the great hall was conveyed, through openings in the vault made of plaster slabs reinforced by iron mesh, into the plenum of the attic which, accumulating heat, acted as an important thermal buffer between interior and exterior. Later replaced by a radiator system, the ventilation system and the plenum above the vaults continue however to function in the same way (Saddy 1977; Leniaud 2002).

1851 is also the year in which the construction of Saint George's Hall in Liverpool ended: this building was widely published in the international literature of the period, and it served as a model for many later applications throughout Europe (Manfredi 2013). It was finished under the direction of architect Charles Robert Cockerell, after the work had begun according to a design by Harvey Lonsdale Elmes (Mackenzie 1864; Sturrock 2017). The hall is the only constructed building of an ambitious renewal program of the city centre, according to which other public buildings facing a square were supposed to be built. All the buildings were designed with a centralised heating and cooling system, planned by chemist David Boswell Reid, who conducted experiments on natural and forced ventilation at the University of Edinburgh. Reid would later be the protagonist of the controversy with Charles Barry regarding the height of the chimneys for the heating system of the Houses of Parliament (see Henrik Schoenefeldt's contribution in this volume).

Similar vicissitudes concerned almost all the public buildings built in that period. The new circular Reading Room of The British Museum (inaugurated in 1857), which Sidney Smirke designed to replace the former one, which had become too small, in close collaboration with Antonio Panizzi, the library's Principal Librarian, was equipped with a low-pressure system built by Messrs Haden (Harris 1998[60]). Also in the previous reading rooms in the Northern wing a high-pressure water system was installed by Angier March Perkins in 1838. It consisted of coils of pipes enclosed in pedestals or radiators, which had the advantage of limiting the convective movements typical of an air system, which would have damaged the books (Harris 1998[61]).

Later, centralised systems were more and more related to ventilation needs, and from the 1870s on, the water systems became predominant over the air systems.

---

[59]P. 51.

[60]Pp. 186, 244.

[61]P. 154. The water systems were proposed at the beginning also by Labrouste for the Bibliothèque Sainte Geneviève exactly for this reason. They were discarded because of the fear of fires, which could cause their explosion, as in the case of Saint-Sulpice Church in Paris in 1858.

There are numerous examples. Among others, a noteworthy example is that of the University of Glasgow in Gilmorehill, whose design was the result of the collaboration between architect George Gilbert Scott and mechanical engineer Wilson Wetherley Phipson. It involved the construction of a ventilating tower, which became "also a sign of the willingness of the new merchant class in town to support the development of new knowledge through research" (Lerum 2016[62]). Another interesting example is the Manchester Town Hall, designed by Alfred Waterhouse with the collaboration of Haden and Son for the mechanical systems, whose spaces were designed also with regard to the flow rates of the air they had to convey: "A hybrid warming system combining the advantages of central hot water heat distribution with the tradition and well-known technology of the open fireplace offered relative simplicity of design and installation while having the ability to handle diverse demands represented by a complex building program" (Lerum 2016[63]).

As we have seen, there is by now enough reference literature on the subject to be able to understand how experimentation on central ventilation and heating systems was part of the shared assets of construction professionals, at least in the European capitals and in the countries which experienced a faster industrial development (Manfredi 2013[64]; Forni 2017).

It is interesting to follow the development of these systems in those buildings in which the hygienic requirements were more urgent, and the connection between wellbeing (lost and to be reinstated) and environment was most evident: hospitals. As in prisons or barracks, the fact that environmental conditions could greatly influence psychophysical conditions and human performances became evident. As Rumford had proved at the end of the eighteenth century, when he was commander-in-chief of staff in Bavaria, a well fed and well-trained army, adequately equipped and healthy, is more efficient than one that must worry about finding firewood and food.

Sanitary structures are among the buildings which underwent the most rapid and explosive changes in the nineteenth century, concerning the internal climate which had to be adapted to the hygienic conditions considered to be best. Also without being based on scientifically verified knowledge in a modern sense, as stated by the English physician Thomas Percival, "Air, diet and medicine, are the three great agents to be employed, in preventing and correcting putrefaction and contagion in hospitals. A gallon of air is consumed every minute by a man in health, a sick person requires a larger supply, because he more quickly contaminates it" (Daniel 2015[65]; Aikin 1771).

[62]P. 89.

[63]P. 101.

[64]Particularly p. 212 and ff., which tries to describe the Italian situation, regarding the fact that there was a great lack of homogeneity between the avantgarde of capital cities and a delay in the applications. The situation in Rome, which evolved significantly only after 1870, when it became the capital of unified Italy, has yet to be studied.

[65]P. 566.

As opposed to an external environment whose deterioration appeared from the start as the price to pay for the mechanisms of progress, it was necessary to counter healthy islands in which one could obtain conditions favourable for social control, education, surveillance and punishment. The buildings built with this objective were, in the first place, hospitals and prisons.

At the end of the nineteenth century, exemplary models were the Royal Victoria Hospital in Belfast, "less progressive in architectural style, but more advanced environmentally" (Banham 1969[66]), and the Birmingham Children's Hospital, equipped with heating and ventilation by propulsion with large centrifugal fans (Lerum 2016[67]). In both, the total lack of correspondence between the severe outer look and the modernity of the technological solutions is displayed "with painful clarity" (Banham 1969[68]). But the debate on heating and ventilation systems began much earlier: in the competition for the Zurich cantonal hospital in the 1830s (Daniel 2015), in the Gebärhaus (maternity ward) in Munich which we discuss later, and then in the construction of the Hôpital Lariboisière in Paris, where a true controversy broke out because of the two technologies which competed for the supremacy of the most efficient and the healthiest (Gallo 2006[69]). Philippe Grouvelle's heating system "mixte vapeur et eau chaude avec une ventilation mécanique mue par la vapeur", and Léon Duvoir-Leblanc's system using hot air and ventilation through aspiration, were used in two different pavilions of the hospital, and the results of the surveys were discussed before a commission presided by General Arthur Morin, a veteran of experimentation in the military cantonments.

At the same time, increasingly greater attention was given to prison buildings, which became another issue on which the institutions of the renewed national states confronted themselves. Pentonville prison, inaugurated in London in 1842, "[…] was visited by the main European heads of state and by the representatives of almost all the governments. The quest for new control and repression technique was by then an international problem" (Dubbini 1986[70]). The heating and ventilation system was designed by a military engineer, Major Joshua Jebb: the building "was a landmark in the technology, and seems to have been adopted by General Morin for many large public buildings in France" (Billington and Roberts 1982[71]). There were many inventions, but one can reasonably affirm that it is difficult, and not especially useful, to establish the supremacy of one method over the other, also because there was a continuous succession of changes. In the case of Pentonville, we have extremely precise data. The required air flow had to be between 50 and 76 mc/h in

---

[66]P. 75 and ff.

[67]P. 123.

[68]P. 83.

[69]P. 270 and ff. The dispute lasted until the 1870s, without revealing a clear supremacy of one system over the other.

[70]Pp. 61–62.

[71]P. 120.

every cell, while the temperature was not to drop lower that 11 °C. The cost of the system was enormous: for a volume served by 56,000 mc, "the installation cost £17,000, and maintenance was £500 a year. It consumed 2.15 tonne of fuel per day" (Billington and Roberts 1982[72]). In the 1854 competition for the Prison Mazas there was opposition again between Duvoir and Grouvelle; a temperature between 13 °C and 16 °C was required for the cells (Gallo 2006[73]). "A sub-committee stated that ventilation was the principal requirement for prison hygiene": ventilation as an obsession?

The issue of the ventilation of enclosed spaces may be tackled from multiple standpoints. In addition to the considerations on the quality of air, which had to be protected from an increasingly more polluted outer environment, as we have seen, the necessary change of air became an aspect which had to be taken into consideration, also due to the rapid dissemination of gas lighting systems, and which determined design, all the way down to the interior details and decoration. Moreover, one should always keep in mind the contamination of air, the infection through the environment, is an element which transcends historical circumstances and is deeply rooted in a very ancient conception of medicine and of science on a larger scale, whose most complete product was the miasma theory. According to this theory, some (almost all) illnesses spread through direct infection, among the people present in a same place (even in the open), if not sufficiently ventilated. Independently from the causes of the various morbose symptoms, which were still largely unknown in the eighteenth century, illnesses such as cholera, typhoid, plague, tuberculosis, were attributed to not better identified miasmas which hovered in the room, whose healthiness would thus have improved through frequent changes of air. These convictions survived in the scientific debate until the late eighteenth century.

The fact that the aetiology of many illnesses was known by then, did not remove the prejudice against the malign effects of polluted air, whatever the causes and amount of the known effects. As stated by Magdalena Daniel, "In those years, prominent hygiene enthusiasts proclaimed that, through a 'health-promoting interior climate', almost any disease could be cured. Despite the fact that, by 1830, it was far from clear which parameters exactly constituted such a climate" (Daniel 2015[74]).

A trial and error method was used. The Gebärhaus in Munich was one of the first hospital buildings constructed with the declared goal of improving the hygienic conditions following a new sensitivity in the design of the spaces. It was built to a design by Arnold von Zenetti in 1854. The sizes of the recovery wards were reduced to the point of obtaining four-bed rooms, since the contagion of puerperal

[72]P. 207.

[73]P. 271. Charles Lucas, the inspector general of the French prisons, wrote as follows: "one must never allow a level of material comfort which exceeds that which the inferior classes may obtain, because one would create, so to speak, a prize encouraging crime". Cited by Dubbini (1986, p. 63).

[74]P. 565.

fevers led to higher infection and mortality rates than those registered among the mothers who gave birth at home. Until the dissemination of Semmelweis's studies, it was believed that the spread of the disease was due to the proximity of the patients and the poor ventilation of the rooms, while it was later discovered that the contagion was caused by repeated contact with the doctors' hands, who did not wash their hands between one visit and the other. It was indeed a problem of hygiene, however not in the expected sense! (Manfredi 2013[75]).

What has been less investigated until now, it seems to me, is the passive behaviour of buildings. One of the functions of architecture is that of supplying an answer to climatic variations. There is a constructive wisdom which has determined the adoption of variable models, which at times have been welcomed and emphasised when acknowledging local or vernacular characteristics in buildings (Olgyay 1963). This aspect has been discussed very little in theoretical elaborations, probably because it was greatly committed to the ability and "know how" of the executors. Since the technical revolution in the nineteenth century, the awareness of the importance of the conditions of comfort has grown rapidly, and has settled on the sharing of standards which at a certain point appeared as optimal for the physical wellbeing of human beings. Pioneering and yet complete studies, such as Baruch Givoni's *Man, Climate and Architecture*, have thoroughly examined issues related to the performance and calculation of the variables which influence environmental parameters, trying to combine psychological aspects with the elements of architectural design (Givoni 1969). This was not only a matter of physiology: Garry Thomson determined parameters for the conservation of works of art, which have been questioned only quite recently (Thomson 1978; Camuffo 1998; Luciani 2013).[76]

But there are very few serious studies which have dealt with this issue in a historical perspective. The reading proposed by Patricia Waddy on the architectural treatises of the Renaissance, from Alberti to Palladio and Scamozzi, which may have influenced the work of later architects, is not systematic, and limits the analysis of the possible mechanisms of climate control to a specific and certainly exceptional moment, that of seventeenth century Papal Rome. The interpretations of Vitruvius depend on the changes of the paradigms and the different conception of science. The understanding of the use of spaces, which varied according to the needs of those who lived in them, is the key element for understanding the environmental strategies: "The most important means for furnishing comfort in palaces was the provision of separate rooms or apartments for the use in different seasons" (Waddy 1990[77]). The flexibility of the rooms is still greatly due to their scarce specialisation, and favours the seasonal use of different apartments: "Rooms with an

[75]Pp. 179–180.

[76]A new reading of the interaction with the environment has been launched, among others, by Camuffo (1998). The reading of the environmental parameters for museums in Luciani (2013) points out the role of the historical climate for preservation.

[77]P. 16. A selection of classic French writers of treatises can be found in Gallo (2006, p. 107 and ff.).

eastern exposure are generally good for spring and autumn, and even for summer, but they do not receive much sunlight during the winter" (Waddy 1990[78]). Occupation depended also on the time of day: "Rooms to be used at particular times of the day may vary the pattern: libraries, for morning use, might enjoy eastern light; rooms for bathing, an afternoon activity for which warmth was desirable, would profit for a western orientation" (Waddy 1990[79]) and so on.

In any case, an analytical study of the environmental contributions of Classical treatises has yet to be written: "Vitruvius gives some advices; writers of the fifteenth, sixteenth and seventeenth century concur; and the practice of architects and inhabitants alike conforms to the pattern" (Waddy 1990[80]).

Dean Hawkes's work, which is more specific, focuses on the environmental behaviour of buildings as an answer to climatic variations, and he identifies the adaptation of Classical models to the British climate made by Palladianism as one of the reasons for its success and longevity. Hawkes, whose aim is "to propose an alternative reading of the link between architecture and climate. In architectural science the method is invariably to represent buildings as a logical response to a pre-existing climate" (Hawkes 2012[81]), shows in a clear way how the variations of architectural construction in the various times and places adapt to the material influences exerted by the environment, in the first place the climatic conditions. Hawkes sees the creation of modern science, which developed over the course of the seventeenth century, as the turning point in the systematic approach to the problem. Christopher Wren, both scientist and architect, epitomizes this situation. The design of the architectural organism of the Sheldonian Theatre in Oxford, among his better-known works, takes on, from this point of view, a meaning which transcends the strictly stylistic and architectural reasons, reaching a balance with the constructive and performance requirements.

More generally, beginning from examples of Elizabethan architecture such as Hardwick Hall or Bolsover Castle, Hawkes proposes a more complex reading, set in the material reality of environmental conditions, to understand how the concepts of distribution of public and private spaces evolved over time. From this specific point of view, also the new churches which dot the urban fabric of London (many by Wren himself and by his pupil Nicholas Hawksmoor), as rebuilt after the Great Fire of September 1666, are significant. The special attention dedicated to the design of their floorplan in relation to the liturgical requirements, determined precise rules: Wren himself "declares the importance of seeing and hearing in the design of a church, hence of daylight and acoustics" (Hawkes 2012[82]). Special attention had to be dedicated also to thermal comfort, which was increased thanks to the presence of wooden box pews: "As built, the churches were unheated and, bearing in mind the

[78]P. 16.

[79]P. 16.

[80]P. 16.

[81]P. 2.

[82]P. 72.

extreme cold of late-seventeenth-century winters, attendance at service would have a thermal ordeal. The enclosure of a box pew protects against draughts and, by conserving the heat of closely and presumably well-clad bodies, provides a slightly more comfortable microclimate" (Hawkes 2012[83]).

Thus, it becomes clear that the *Vitruvius Britannicus* is not only a translation of the ancient text. A true analysis of Colen Campbell's work "[…] suggests that, in matters of response to climate, Campbell followed British precedents, rather than the unmodified Italian formulae of Palladio" (Hawkes 2012[84]).

Some attempts at analysis have been carried out regarding the protection of interior spaces in hot climates, or in the hot seasons of temperate areas. The defence from high temperatures in summer was traditionally greatly entrusted to the thermal inertia supplied by the generous thickness of old walls.

This expedient, which was supported by constructive needs, has always been accompanied by efficient use of the interior spaces, aimed at minimising the variations. The few surviving documents concerning the buildings built by the Arab-Norman civilisation confirm this. The "paleo-technology"[85] analysed in the volume *I sistemi di ventilazione naturale negli edifici storici* shows how the custom of using flows of cool air coming from the subsoil, already described by Hero of Alexandria and Pliny the Younger, continued also in more recent times. The cases of the Trenti villas in Costozza di Longàre, near Vicenza, are known. It is a group of buildings built and then remodelled over time between the fifteenth and the nineteenth century, built above a group of underground caves from which air flows naturally; due to the thermal exchange with rock in the depths, this air flow is constantly at a temperature which is much cooler that the outer air in summer, while in winter it succeeds in mitigating the cold weather. This natural phenomenon has been tamed and curbed to some extent, and numerous ventiducts, provided with mechanisms for the regulation of the flows, have been built to serve inhabited environments (Hawkes 2012[86]).

More focused studies (although still very general in character) have recently been dedicated to nymphaeums, crypto porticos and Sirocco chambers, which made up the rooms functionally specialised to offer a cool climate also in the very hot summer days in southern Italy (Minutoli 2009). The presence of water, which was left running as a trickle in the style of the Arab irrigation systems, contributed to the raising of the level of humidity and to the absorption of ambient heat, as in the Zisa in Palermo or in the Alhambra in Granada.

Pre-industrial age buildings reveal the knowledge of this ancient wisdom. Palazzo Corsini and later Palazzo Pitti in Firenze, and Palazzo Marchese in Palermo, show how conscious the design choices were, since the orientation and

[83]P. 75.

[84]P. 97.

[85]From the *Presentazione* of Francesco Gurrieri to Balocco et al. (2009, p. 5).

[86]This system is described with suprise by Palladio in his *Quattro Libri dell'Architettura*, 1570, Chapter XXVII.

distribution of the rooms were closely tied, in such a way as to make the most of the temperature gradients which formed naturally between the rooms, with the aim of establishing convective flows which could reduce the oppressive heat (Balocco et al. 2009).

The inexhaustible construction process of the Alhambra in Granada supplies many hints on this. The in-depth study dedicated to it by Todd Willmert, reveals "a wealth of passive strategies to cope with climate, providing insight into how the palaces were inhabited and the lifestyle they reflect" (Willmert 2010[87]).

In the extreme climate conditions of the plateau of Granada, many different uses were made of the complex, among which the decisive ones tied to the Arab conception, with the construction of the Comares and Lions Palaces during the dominion of the Nasrids in the fourteenth century, and later the construction of Charles V's Renaissance palace in the first half of the sixteenth century, which provided a solution to the needs for comfort of a larger and more demanding royal court. One can deduce that "despite cultural differences, both the Nasrids and those who succeeded them used, in the broadest sense, similar strategies in designing and inhabiting spaces, although they applied them in profoundly different ways" (Willmert 2010[88]).

It seems that the adjustment to the climate followed dissemination routes which went well beyond cultural identity, and spread rapidly in very distant contexts also: "Indeed, the same strategies exploited by the Nasrids to create a habitable architecture were advocated in non-Muslim treatises and manifest in architecture from Roman times to the Renaissance" (Willmert 2010[89]; Waddy 1990[90]).

Within the complex, layered architectural organism, strategies aimed at opposing the extreme climate conditions (from −13 °C in winter to +45 °C in summer) were implemented. In particular, the buildings were massive, with openings designed to favour ventilation and to contrast the violent sunlight, typical of that latitude. In these buildings a fundamental role was played by running water in the courtyards and also in the rooms, which favoured the processes of evaporative and evapo-transpirative cooling; the orientation and the form of the rooms and courtyards, with high and narrow spaces, limited-sized facades facing south and protected by porticos to protect from the sun in the summer and to allow sunlight in winter; the presence of insulating spatial buffers, the differentiated use of the rooms —an aspect of which little is known, since there was the custom of changing the use according to the arrangement of movable furniture. All these aspects caused substantial variations of the microclimatic conditions (Willmert 2010[91]).

[87]P. 157.

[88]P. 158.

[89]P. 172.

[90]Chapter 2, *Comfort*.

[91]Willmert has also surveyed, by using instruments, the environmental conditions in summer and winter.

I think I have shown how the vicissitudes related to the equipment of mechanical systems, which were by no means marginal in determining the trends in the practice of building, have been relegated to secondary spaces in architectural literature.

The climate, the variations which conditioned it even in recorded history, and the adjustment techniques man has searched for to build well-adapted buildings, are the subject of this volume. But much has yet to be investigated. The renewed frequency of climatic changes, the fear of their irreversibility with regard to the conditions we are familiar with, have allowed us to become aware of the fact that an approach which aims only at manipulating the external conditions is no longer effective, and furthermore, ill-adapted to our time. A different sensitivity will be useful for defining in contemporary terms how certain relationships between man and environment were established in the past (this relationship has escaped analysis for a long time), both for the understanding of historic buildings and for their protection, and for the aware design of objects and of the reciprocal interrelations we shall establish in future research.

References

Agamben G (2009) "What is an apparatus?" And other essays. Stanford University Press, Stanford, pp 1–24

Aikin J (1771) Thoughts on hospitals. London

Bachelard G (1938) La psychanalyse du feu. Paris

Balocco C, Farneti F, Minutoli G (eds) (2009) I sistemi di ventilazione naturale negli edifici storici: Palazzo Pitti a Firenze e palazzo Marchese a Palermo. Alinea, Firenze

Banham R (1969) The architecture of the well-tempered environment. The University of Chicago Press, Chicago

Billington NS, Roberts B (1982) Building services engineering: a review of its development. Pergamon, Oxford

Brimblecombe P (1987) The big smoke. History of air pollution in London since medieval times. Routledge, London and New York

Brimblecombe P (2010) Air pollution and society. EPJ Web Conf 9:227–232

Bruegmann R (1978) Central heating and forced ventilation: origins and effects on architectural design. J Soc Arch Hist 37(3):143–160

Campbell J, Papavasileiou S, Makrodimitri M (2012) The construction and integration of historic heating systems in churches in the United Kingdom from the 17th to the early 20th century. In: Nuts and bolts, proceedings of the 4th ICCH, Paris, pp 277–288

Camuffo D (1998) Microlimate for cultural heritage. Elsevier, Amsterdam

Cook J, Hinchcliffe T (1995) Designing the well-tempered institution of 1873. Arch Res Q 1 (2):70–78

Cook J, Hinchcliffe T (1996) Delivering the well-tempered institution of 1873. Arch Res Q 2 (1):66–75

Cruickshank D (ed) (1985) Timeless architecture: 1. Architectural Press, London

Daniel M (2015) Constructing health—the pursuit of engineering a "health-promoting interior climate" during the 1830s and 1840s. In: Proceedings of the 5th international congress on construction history, vol I, Chicago, 3–7 June 2015. Chicago, pp 565–572

Davies C (1985) Dulwich picture gallery (ed: Cruickshank D), pp 69–88

Dubbini R (1986) Architettura delle prigioni: i luoghi e il tempo della punizione, 1700–1880. F. Angeli, Milano

Eleb-Vidal M, Debarre-Blanchard A (1989) Architectures de la vie privée. Maisons et mentalités XVIIe-XIXe siècle. Archives d'Architecture Moderne, Bruxelles

Elliott CD (1992) Technics and architecture. The development of materials and systems for building. Cambridge, London

Forni M (2017) La «stufa alla moscovita» a Milano: applicazioni di un sistema di riscaldamento ad aria calda nei secoli XVII e XVIII (ed: Manfredi C), pp 58–111

Gallo E (2006) Modernité technique et valeur d'usage: le chauffage des bâtiments d'habitation en France. Unpublished PhD dissertation, Université Paris I Panthéon Sorbonne, UFR Histoire de l'art et archéologie

Givoni B (1969) Man, climate and architecture. Elsevier, Amsterdam

Guillerme A (1983) Le temps de l'eau. La cité, l'eau et les techniques. Champ Vallon, Seyssel

Guillerme A (2007) La naissance de l'industrie à Paris entre sueurs et vapeurs. Champ Vallon, Paris

Halliday J (1999) The great stink of London. Sir Joseph Bazalgette and the cleansing of the Victorian metropolis. Sutton Publishing, Stroud

Harris PR (1998) A history of the British museum library 1753–1973. British Library, London

Hawkes D (1985) St. Andrew's Roker (ed: Cruickshank D), pp 8–22

Hawkes D (2008) The environmental imagination. Technics and poetics of the architectural environment. Taylor and Francis, London

Hawkes D (2012) Architecture and climate. An environmental history of British architecture. Routledge, London and New York

Huber A (2012) Ökosystem Museum—Ein konservatorisches Betriebskonzept für die Neue Burg in Wien. In: Nachrichten der Initiative Denkmalschutz, Nr. 11, Juni–Sept 2012, p 27

Leniaud J-M (sous la direction de) (2002) Des palais pour les livres: Labrouste, Sainte Geneviève et les bibliothèques. Maisonneuve et Larose, Paris

Lerum V (2016) Sustainable building design. Learning from nineteenth-century innovations. Routledge, London and New York

Loyer F (1987) Paris XIX siècle. L'immeuble et la rue. Hazan, Paris

Loyer F, Delhaye J (1986) Victor Horta, Hotel Tassel: 1893–1895. Archives d'Architecture Moderne, Bruxelles

Luciani A (2013) Historical climates and conservation environments. Historical perspectives on climate control strategies within museums and heritage buildings. Doctoral dissertation, Politecnico di Milano

Mackenzie W (1864) On the mechanical ventilation and warming of St. George's Hall, Liverpool. Civ Eng Arch J, XXVII

Manfredi C (2013) La scoperta dell'acqua calda: Nascita ed evoluzione dei sistemi di riscaldamento centrale. Maggioli, Sant'Arcangelo di Romagna

Manfredi C (ed) (2017) Architettura e impianti termici. Soluzioni per il clima interno fra XVIII e XIX secolo. Allemandi, Torino

Minutoli G (2009) Tecniche di ventilazione naturale nell'edilizia storica (ed: Balocco C et al), pp 9–39

Olgyay V (1963) Design with climate: bioclimatic approach to architectural regionalism. Princeton University Press, Princeton

Olley J (1985) The reform club (ed: Cruickshank D), pp 23–46

Olley J, Wilson C (1985) The natural history museum (ed: Cruickshank D), pp 47–67

Parrie E, Dernie D (1985) Hôtel Tassel, Brussels (ed: Cruickshank D), pp 89–108

Perrot M (1989) Histoire de chambres. Seuil, Paris

Richardson CJ (1837) A popular treatise on the warming and ventilation of buildings. Weale, London

Saddy P (1977) Henri Labrouste architecte 1801–1875. Caisse nationale des monuments historiques et des sites, Paris

Schott D, Luckin B, Massard-Guilbaud G (eds) (2005) Resources of the city: contributions to an environmental history of modern Europe. Ashgate, Aldershot

Sori E (2001) La città e i rifiuti: ecologia urbana dal Medioevo al primo Novecento. Il Mulino, Bologna

"State and Church" (1858) The Examiner, 3 July 1858

Sturrock N (2017) St. George's Hall, Liverpool—a major refurbishment and a new heritage centre for the world's first air-conditioned building (ed: Manfredi C), pp 227–243

Thomson G (1978) The museum environment. London

Vigarello G (1985) Le propre et le sale: l'hygiène du corps depuis le Moyen Âge. Seuil, Paris

Waddy P (1990) Seventeenth-century Roman palaces: use and the art of the plan. Architectural History Foundation, New York

Willmert T (1993) Heating methods and their impact on soane's work: Lincoln's inn fields and Dulwich picture gallery. J Soc Archit Hist 52(1):26–58

Willmert T (2010) Alhambra palace architecture: an environmental consideration of its inhabitation. In: Necipoğlu G, Leal KA (eds) Muqarnas. An annual on the visual cultures of the Islamic World, no 27. Leiden and Boston, pp 157–188

Wright L (1964) Home fires burning. The history of domestic heating and cooking. Routledge and Kegan Paul, London

Wright L (2005) Clean and decent: the fascinating history of the bathroom and the water-closet. Penguin Global, London

Heating Verona in the Nineteenth Century. From the Fireplace to the Hot Water Systems

Marco Cofani and Lino Vittorio Bozzetto

Abstract Verona, a city of significant importance for Northern Italy, lived several administrative seasons between the end of the eighteenth and the beginning of the twentieth century, passing from the ancient regime of the Serenissima Republic of Venice to the Kingdom of Italy, through French and Habsburg domination. In this long period, the city has repeatedly changed its territorial role and urban structure, moving from an apparently immutable condition to an impressive architectural and building development. In this sense, the new buildings but also the historical ones were invested by profound changes from the point of view of technological equipment, first of all those relating to heating systems. New aspects such as air quality, comfort and safety became central. The changes affected both the private and public construction sectors, which gave a decisive push towards the adoption of new, more efficient and easily managed plants, especially inside the most modern building types such as schools, hospitals and public offices.

This paper, although it is the result of reflections and research common to the two authors, was written by Marco Cofani as regards Sects. 1–3 and 5–7, and by Lino Vittorio Bozzetto as regards Sect. 4.

M. Cofani (✉)
Ministero per i Beni e le Attività Culturali (MiBAC), Soprintendenza Archeologia, Belle Arti e Paesaggio per le Province di Verona, Rovigo e Vicenza, Verona, Italy
e-mail: marco.cofani@mail.polimi.it

L. V. Bozzetto
Verona, Italy

C. Manfredi (ed.), *Addressing the Climate in Modern Age's Construction History*,
https://doi.org/10.1007/978-3-030-04465-7_2

1 Verona, 19th Century: Between Geopolitical and Military Centrality and Industrial Backwardness

From the end of the eighteenth century (Berengo 2009[1]) and at least until the annexation of Veneto to the Kingdom of Italy (1866), the city of Verona played a major role in the geopolitical and military chessboard of northern Italy, covering the most important garrison of the Hapsburg Quadrilateral in Lombardo-Veneto. Its economy, historically based on agriculture and trade, at that time experienced a strong impulse in the construction sector, especially after 1834 with the field marshal Joseph Radetzky as commander of the Hapsburg army in Italic territory. Radetzky, in fact, took the decision to modernize and expand the ancient but at the same time powerful Scaliger and Venetian fortifications that already defended the city along the Adige river (Tonetti 1997; Preto 2000).[2] In fact, it must be remembered that, since 2000, Verona is part of the UNESCO World Heritage List due to its two-thousand-year fortification and military architecture tradition, from the Roman period to the Habsburg domination.

If the military role of Verona was central, not so for the productive sector. Until the mid-nineteenth century, the city—which could count on an already very backward industrialization, as well as the internal northeastern territory of the italic peninsula—saw a further stagnation in this sector. This was because a large part of the Austrian administration's economic efforts was directed to the fortification sector and to the maintenance of the huge military garrison (Selvafolta 1994). The defensive strategy of the city, moreover, provided in case of siege the imperative need not to obscure the shooting towards the campaign of the artillery, positioned along the fortified walls and fort. From the point of view of urban development, this constraint imposed the absolute prohibition of building large factories outside the walls, where extensive plots of land were instead expropriated and reduced to military servitude.

It was only in the second half of the 40s that some new industries, although still small, began to lay the foundations of the future development, whose effect on the building technology sector will only be seen after the Unification of Italy.

The first industrial plant of a certain importance—studded by very serious inefficiencies and malfunctions in the first years of activity—was the gasometer, built in 1845 near the monumental cemetery, in the south part of the city. Here for some decades the fossil coal was distilled to produce the gas needed to power the city's public lighting. The contract for the construction of the gasometer was stipulated with the Parisian company "Franquet & Latruffe" (Landi 2010[3]; Ferrari 2012[4]), while the raw material—the fossil coal—came from Wales through the

---

[1]Pp. 88–113.

[2]About the Hapsburg society and economy in the Veneto region, see Tonetti (1997) and Preto (2000).

[3]P. 126.

[4]P. 185.

Venetian port, connected to the mainland by the newly inaugurated *Ferdinandea* railway (Bernardello 1996; Cattaneo and Milani 2001).

The construction of the railway line from Milan to Venice and the inauguration, in 1849, of the veronese station of Porta Vescovo—strictly functional to the military sector of the city—were decisive for the industrial development and the establishment of some factories, including the company for the construction of railway wagons (Officine Ferroviarie Veronesi), the cotton mill of Montorio (founded in 1847) and the same gasometer: this was due to the significant reduction of the coal transportation costs, whose trade started to become sustainable in spite of the still heavy duties imposed by the Austrians. The gasometer, for example, was expanded as early as the 1850s, with as many as 5 gas meters, a distribution network of around 30 km and a consumption of fossil coal of 4500 tons per year (Landi 2010).

2 The Problem of Energy Sources

Around the middle of the century, in Verona, the use of fossil coal was not yet extended to the sector of building heating, which would have allowed to exploit the new and improved technologies for the production and distribution of heat already present in other Italian contexts, as for example in Lombardy and Piedmont. And this despite, for some decades, one of the main topics of study, in the Verona area, was the new fuels alternative to wood for heating buildings. In the first half of the nineteenth century there were, in fact, several remarkable attempts by eminent Veronese scholars—all belonging to the prestigious Academy of Agriculture—to trace, within the Verona area and neighboring ones, the presence of fossil fuels with greater calorific value than wood, able to provide a partial energy independence to the city. In particular, the essay by Count Ignazio Bevilacqua Lazise on fossil fuels in the province of Verona, dated 1816, in which the noble scholar made—with the help of the chemist Giacomo Bertoncelli—a remarkable survey in the mountainous areas of the province in search of the *litantrace*, a good quality coal (Bevilacqua Lazise 1816). The Count had a dual objective: to solve the perennial scarcity of wood, which was become extremely expensive also because it was largely imported from the Tyrol; to stop the consequent and increasingly marked destruction of the mountain forests of the province, which were causing huge problems of hydrogeological instability.

The Count's study led to the discovery of two important carboniferous sites, one in the village of Bolca in the municipality of Vestena Nuova, in the Lessini mountains, and the other in Chiampo, not far away but already in the Vicenza area. Both sites were effectively exploited for a short time, especially in the 30s and 40s, but they were never able to bring the great benefits assumed by the count and also by other scholars who after him took an interest in these topics (Bertoncelli 1838; Da Lisca 1841; Scopoli 1841). This was due not so much because of the poor quality of the fossil fuel, which according to Bertoncelli's analysis was comparable to the English one, but essentially to the lack of adequate investments, technical means and infrastructures able to allow the excavation and transportation of adequate quantities of material and the creation of a real industrial sector.

A further limit to the use of fossil coal for heating buildings was imposed, this time for political reasons, after the second war of Italian independence (1859), when the restrictive measures imposed by the Hapsburg government led to a sharp reduction in imports and consumption (Ferrari 2012[5]).

It is therefore possible to affirm that in the Verona area, at least until the first years after the annexation of the Veneto to the Kingdom of Italy, the technologies for heating buildings were substantially limited to the chimneys and traditional stoves, in metal or majolica, even if these—as we shall see—were sometimes used in a rather innovative way, especially in the new military buildings. Regarding the fuel, it was mainly used charcoal, wood, mostly imported from the forests of South Tyrol, and, more rarely, fossil coal.

3 The "Venetian" Heating Systems. The Tradition of the Fireplace Between Functionality and Representation

In the Verona aristocratic residences, even after the mid-nineteenth century, the chimneys still represented the main and most widespread heating system. Their importance in the rooms of the noble palaces was clearly not only linked to the comfort and the production of heat, but also—and in some cases above all—to representation of the rank and the wealth of the family. The fireplaces were in fact precious objects, made with noble materials, often finely carved or decorated. The survey on the veronese archives has so far allowed to trace only few cases, however of considerable interest, in which in addition to the description of the interventions, the drawings of the fireplaces were also found with a lot of decorative details.

A first example is the three fireplaces commissioned in 1788 to Pietro Puttini by the Marquis Alessandro Carlotti for his magnificent palace, in front of the Roman *Porta dei Borsari*. Two of the fireplaces were of identical design but made one in brick and the other in Carrara marble (Fig. 1a); the third was made in a local yellow limestone (*pietra Gallina*) with the decorations by the famous artist Cignaroli (Fig. 1b). The drawing that represents the first two is quoted and shows a net size of the hearth of four veronese feet (about 137 cm). Together with the drawings of the fireplaces, among the Carlotti's documents it has been also found a perspective drawing of a room entirely decorated, both on the walls and on the vaulted ceiling with late-Baroque motifs, in which an elegant fireplace appears at the center of the back wall, surmounted by a precious clock and a mirror (Fig. 2): it is not known either the author or what exactly it is, if an idea for the preparation of a salon inside the family palace or a simple architectural suggestion of the Marquis taken from some publication.

[5]P. 127.

Fig. 1 Project of Carlotti's family fireplaces

Fig. 2 The Carlotti's perspective drawing with a decorated fireplace in the centre

A second example—probably dating to the early nineteenth century even if the drawings in this case do not bear any reference—concerns some graphic representations of fireplaces preserved in the Del Bene family archive (Fig. 3[6]), owners of a palace in the city and of a large villa in the town of Volargne, a small town on the bank of the Adige river, north of Verona. They are neoclassical fireplaces, very refined and designed by a sure and expert hand, where in two cases the artefacts are quoted and represented in plan and also in elevation, from which it is possible to fully understand both the artistic and functional aspects linked to the position and the size of the fireplace. Both chimneys have a net size of the heart of three veronese feet (about 103 cm) and are particularly interesting because the first one hypothesizes a double stylistic solution for the façade while the second, much more articulated, has two corbels in place of the lateral shoulders, above which are supported two caryatid busts which in turn hold the architrave.

Although still too limited to offer some generally valid considerations, these initial investigations on heating systems in the Veronese aristocratic buildings between the eighteenth and nineteenth centuries highlight, as we said earlier, the still vast and well-established diffusion of chimneys, whose dimensions seem to gradually shrink, probably to ensure greater efficiency in heating and lower fuel consumption. This may also have been driven by the introduction of the small and

[6]ASVr, Famiglia del Bene, mappe e disegni, 39/6, 50, 52, 53, 54, drawings with no date.

Fig. 3 The drawings of two decorated fireplaces in the Del Bene family archive

very efficient Franklin stove, whose model at the end of the eighteenth century had inspired the production of new brick stoves in the Reasso factory in Castellamonte, Turin (Scalva 2010[7]), which spread rapidly throughout northern Italy.

4 The "Hapsburg" Heating Systems. Innovative Use of Traditional Stoves in Veronese Military Buildings

To assess the changes in the heating systems used in Verona during the nineteenth century, it is essential to focus now on the military buildings built by the engineers of the Hapsburg Empire. As it is well known, these buildings represent the extraordinary union of the Veronese constructive tradition with the Central European military architecture, distinguished by a great attention to the problems of comfort and health, as well as, of course, solidity and strength. For these goals, the military engineers adopted particularly innovative solutions: about heating systems, these solutions were based on the simultaneous use of many high efficiency stoves, whose installation was uniformly replicated within different types of buildings, especially barracks and lodgings for officers. Some very interesting examples, partly still preserved, can be found in the buildings of Peschiera del Garda (Bozzetto 1997), another Veronese *piazzaforte* (garrison) of the Quadrilateral (Fig. 4), modernized by the imperial army in the years following 1848 and, from 2017, part of the UNESCO World Heritage List.

Kaiserlich und königlich: more than severely martial, it is Imperial and Royal (*k.u.k.*) the figurative and architectural source of the green square of the Military Districts in the fortress of Peschiera, with its four neoclassic buildings to surround the large park, according to canons of proportion, balance, symmetry and perspective. Felix von Stregen's[8] 1817 project reformed the fortified architecture of the sixteenth-century, from the Porta Verona walls to the bank of the Mincio (*Canale di Mezzo*). The new urban layout took shape progressively during a long-planned building cycle: in 1822 the *Franciscus I* Infantry Barracks, in 1854 the *Palazzo del Comando*, in 1856 the Officers' Pavillion, and finally the Artillery powder building and barracks in 1857 (Bozzetto 1997[9]) (Fig. 5).

The architectural style, essential and thrifty, is influenced by the rational eighteenth-century Hapsburg classicism, devoid of any excess. A Viennese aura also pervades the interior space of these architectures. All this ineffable imprint is represented, in Peschiera, in the *Palazzo del Comando* (*Festungscommando-Gebäude* or *Commandanten-Haus*), the perspective summit of the tree-lined square and the field marshal Joseph Radetzky's headquarters on Garda Lake (Figs. 6 and 7). On the

[7]It was a terracotta fireplace with air circulation and visible fire designed according to Benjamin Franklin's studies and reproduced in Italy in the laboratory of Pietro Reasso of Castellamonte.
[8]Felix von Stregen was an officer of the Hapsburg *Ingenieur-Corps*.
[9]Pp. 185–261.

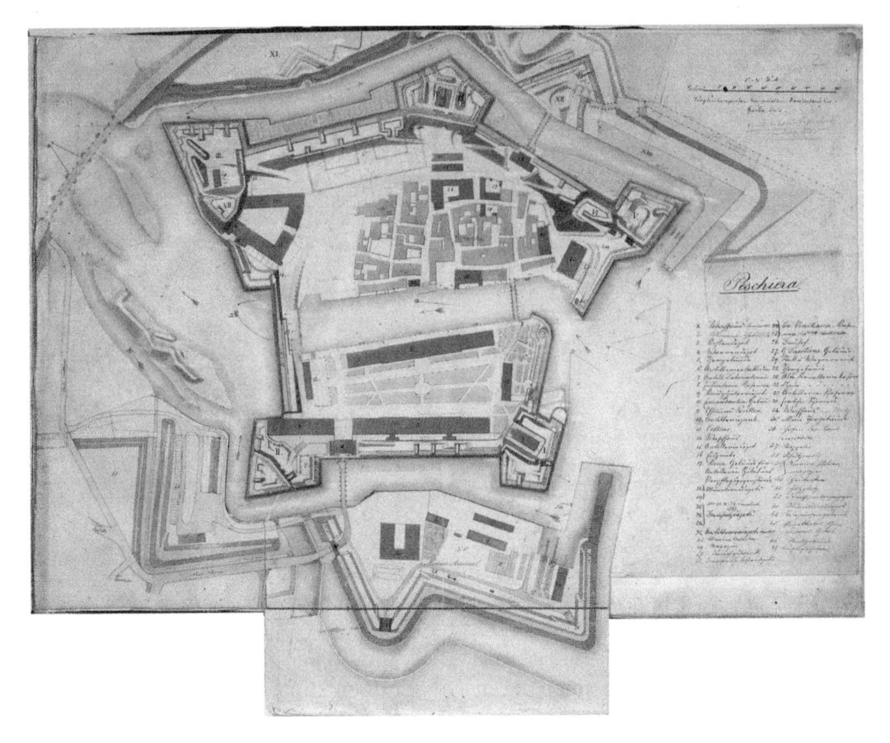

Fig. 4 The piazzaforte of Peschiera del Garda in 1865

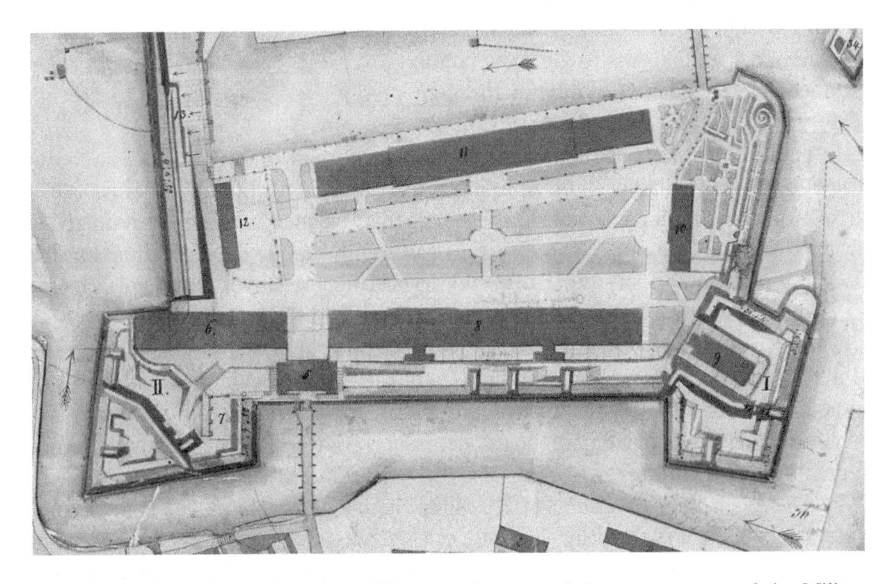

Fig. 5 A detail of the previous map. We recognize, around the green square of the Military District, the Palazzo del Comando (n. 10), the Officers' Pavillion (n. 11), the infantry (n. 8) and artillery (n. 6) barracks and the Venetian Porta Verona (n. 5), connected to the drawbridge

Fig. 6 The Palazzo del Comando of Peschiera in a postcard from the early twentieth century

first floor, inside the two main rooms of representation and reception (Fig. 8), overlooking the park, the stoves covered with white glazed majolica draw our attention: they are two notable *Kunstofen*, or artistic stove, of the middle of the nineteenth century. Original and unmistakable, sometimes bizarre in their constitutive and stylistic evidence, these stoves represented a technological equipment characteristic of the Austro-German culture: over the centuries, the *Kunstofen* assumed a preponderant symbolic meaning of domestic ease, in the nineteenth century widespread among the bourgeoisie.

In addition to the construction methods of the stoves, particular care and constructive rationality had been taken to insert the stoves in the building rooms, as can be seen in the position of each heating elements, isolated or grouped around the corners of a wall crossing. In the second case, with the advantage of making the loading, cleaning and any other operation related to the stoves independent: they were, in fact, accessible from neighboring service rooms.

The artistic stove was imposed in the architectural interior as a mysterious, beneficial presence of fine figure, secretly animated by the fire, silent and guarded, without emission of ashes or fumes. On the roof stood the towers of the chimneys, established as dominant emergencies, necessary complement to the overall configuration of the building.

In the *Palazzo del Comando* of Peschiera the meeting with stoves is now completely unexpected: the two survived stoves (Fig. 9) with a rectangular plan, of high stature, resting on a closed pedestal, surely belong to the original fixed decoration of the building, coeval to the building completed in 1854. They are proportioned, also figuratively, to the size of the rooms, rotated diagonally (at about 45°) and located in

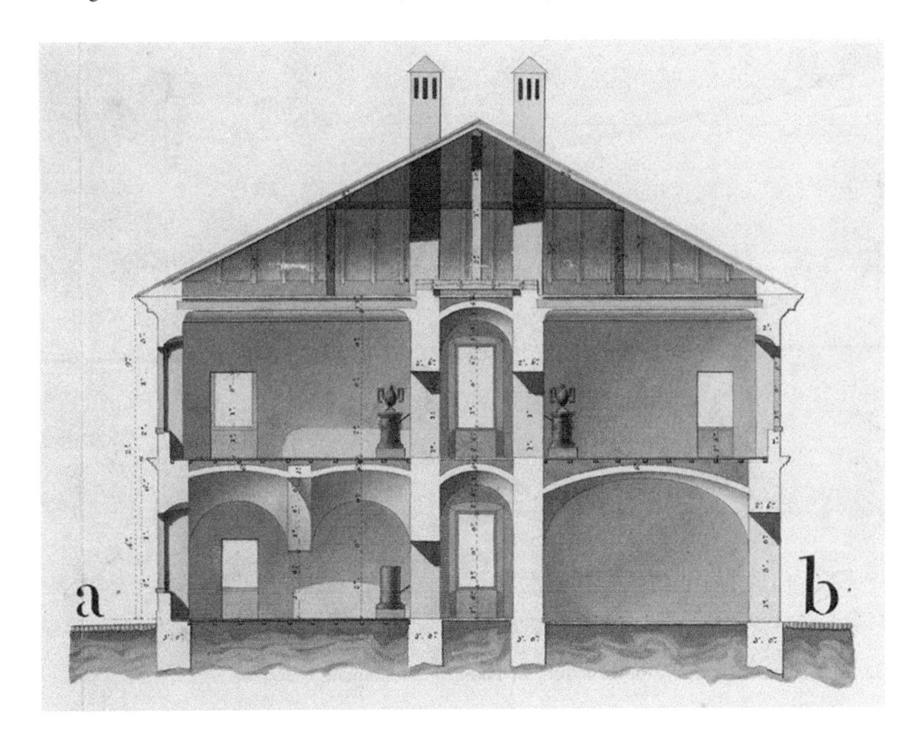

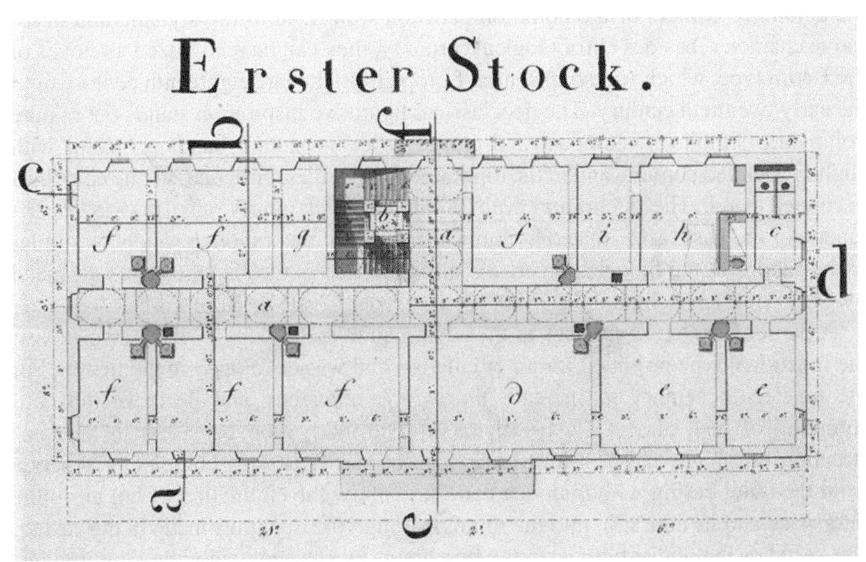

Fig. 7 The original project (1817) of the Palazzo del Comando of Peschiera, completed in 1854 with some changes, in particular to the type of stoves used. Vertical section and plan of the first floor

Fig. 8 The main room on the first floor of Palazzo del Comando

the adjoining corners of the common partition wall. Due to the stylistic and decorative character, besides chronological certainty, they can be recognized as stoves of the Berlin type, which spread in central Europe from the late eighteenth century until the early twentieth century. The neoclassical figurative inspiration stands out in pure volumetric simplicity, compact and closed, with horizontal bands, graceful with slightly rounded corners, and in the top frame with ovules, enriched, in one case, by a decorated band. The decoration with floral bas-relief motifs, also inserted in the lunette of the base, accentuated by the white finish of the majolica panels, is similar to the canons of the first classicism, already proposed in the elegant stoves designed by Carl Friedrich Schinkel, in 1820, for the Tegel Castle in Berlin.

A particular interest also lies in the construction technology of these two stoves: the functional type provided for air circulation and was developed in the first half of the nineteenth century to increase the caloric efficiency and decrease the consumption of fuel, almost always charcoal. The innovation consisted in the construction of an internal cast-iron body for the burner; it was completely separate from the outer casing with majolica panels, to allow the circulation of hot air in the interspace and its emission into the rooms, through the openings made in the casing. The metal or majolica gratings, in the Peschiera stoves decorated with plant weaves, reveal the original Berlin technology for the circulation of hot air.

These stoves also allow to exemplify the original criterion, dictated by constructive needs and practical functionality, to group the stoves in the adjoining corners of a wall crossing: in this way, two or three stoves could be served through

Fig. 9 The two survived kunstofen of the Palazzo del Comando

Fig. 10 The niche behind the stoves in the Palazzo del Comando

the doors arranged in an open niche located the fourth angle (Fig. 10). This allowed a saving of space in the rooms and a rational concentration of the flues. The position of the stoves had to respect pre-established geometrical and dimensional outline of angulation and distance from the walls to guarantee convenient access to the service doors in the niche to load the fuel, start or regulate the fire of each stove and remove

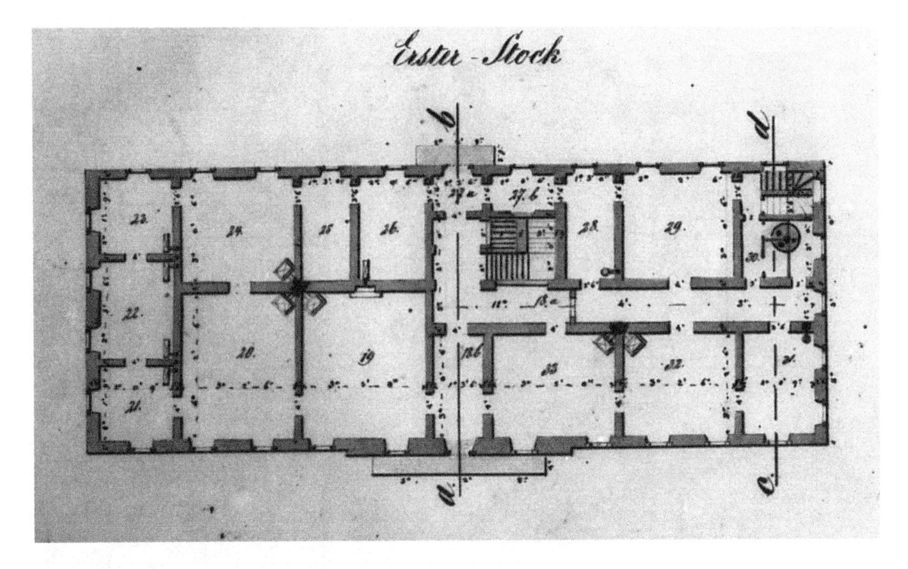

Fig. 11 Palazzo del Comando, 1857 survey, first floor. Room n. 19 is the one photographed in Fig. 8

the ashes, all without disturbing the military officers or commanders. It was therefore a system that, if it still could not be properly called centralized, summarized some of the concepts that we will see applied in the Verona area only after the Unification of Italy.

The original drawings of the building allow us to distinguish and study the complete equipment of the heating appliances, rightly represented as fixed elements of a technological system. In the survey of 1857 (Fig. 11) we can recognize, in the main rooms of the first floor, a system of three large quadrangular stoves and another of two stoves with their canonical arrangement, in the corner of the wall crossing. In the left wing, first floor, there were stoves of different shapes in the smaller rooms, with a very long rectangular plan, leaning against the wall and therefore less cumbersome (Fig. 12a). They could be multi-elements cast-iron stoves called *Etagenofen*, because they were built by overlapping similar prismatic elements in height. The ingenious construction type, developed in the early nineteenth century, introduced a significant technical progress in heating due to internal serpentine piping, which forced the combustion gases to a longer path, resulting in a greater caloric transfer.

Cylindrical cast-iron stoves (Fig. 12b, c) are depicted throughout the ground floor level and in some rooms on the upper floor. This further construction type, very common in nineteenth-century bourgeois homes, is due to the early eighteenth century prototypes of cylindrical stoves, made in cast iron, iron or terracotta.

Fig. 12 The early eighteenth century rectangular and cylindrical cast-iron stoves

Their characteristic dimensions were commensurate with the volume of the rooms.[10] In the original project of the *Palazzo del Comando* (1817, Fig. 7a), then implemented with some variations (1854), there were some characteristic cylindrical stoves, certainly in cast iron, with the base closed and crowned—only those on the upper floor—by an ornamental vase in rococo style.

Today in the main room, in the middle of the back wall in front of the windows, a classical style fireplace also stands, made with polished white marble (Fig. 7b); it dialogues figuratively with the nearby white majolica *Kunstofen* and adds a consonant note of elegance to the room. The two surviving stoves also reveal, in the cracks and in the varnish fractures, the secret suffering of all things in the passage of time. Although they have been extinguished for many decades, there is still a domestic *Biedermeier* warmth, as suited to the noble palace of the Hapsburg Command, of parsimonious, classical imperial beauty.

Changing building type, the Hapsburg barracks for the soldiers' quarters were very large buildings, made by replicating a series of modular rectangular rooms,

[10]Referred to cylindrical terracotta stoves, the following parameters were estimated from experience: for rooms of 10 square *Klafter* (36 m^2) a stove with a diameter of 48 cm, 1.28 m high, was needed; for large rooms of 24 square *Klafter* (86 m^2) the diameter of 65 cm was prescribed for the height of 1.44 m. The cylindrical stoves of cast iron had smaller measures: the diameter ranged from 32 to 54 cm, the height from 80 cm to 1.28 m. While admitting that, with the same overall dimensions and height, the square stoves were equipped with a larger radiating surface, and therefore more efficient; the cylindrical ones was preferred for reasons of aesthetic and for the absence of edges, easily damaged.

generally barrel-vaulted and equipped with two large windows. The rooms were connected to each other by a long corridor, usually located along one of the main façades but sometimes even in the middle of the building. Each modular dormitory owned a cast iron stove, placed in the center of the wall that separated two rooms, frontally powered by wood and/or charcoal and connected to a flue inserted into the wall. The stove next door, put on the other side of the wall, had another flue placed next to the previous one; both flues ended in a single chimney. Some examples of this structure are the barracks of artillery and infantry of Peschiera and the Campone barrack of Verona (Perbellini and Bozzetto 1990[11]).

In other cases, as in the lodgings of the officers—typologically more complex and equipped with more space and comfort—the cylindrical cast iron stoves were placed in the center of the house and organized in a much more articulated manner, as for the *Palazzo del Comando* of Peschiera. These lodgings were also inserted in large buildings and distributed like the barracks by a side corridor: the single lodging, which also housed the family of the officer and the service staff, generally used two building modules divided into different sub-modular rooms. It is the case of the magnificent Peschiera Officers' Pavilion (Fig. 13), built along the bank of the Mincio river inside the fortress. Here, many cylindrical cast-iron stoves are represented in the longitudinal section of the original project of 1817 (Fig. 14), then built in 1856 (Fig. 15). Also in this case the stoves, supported by shaped metal legs, was

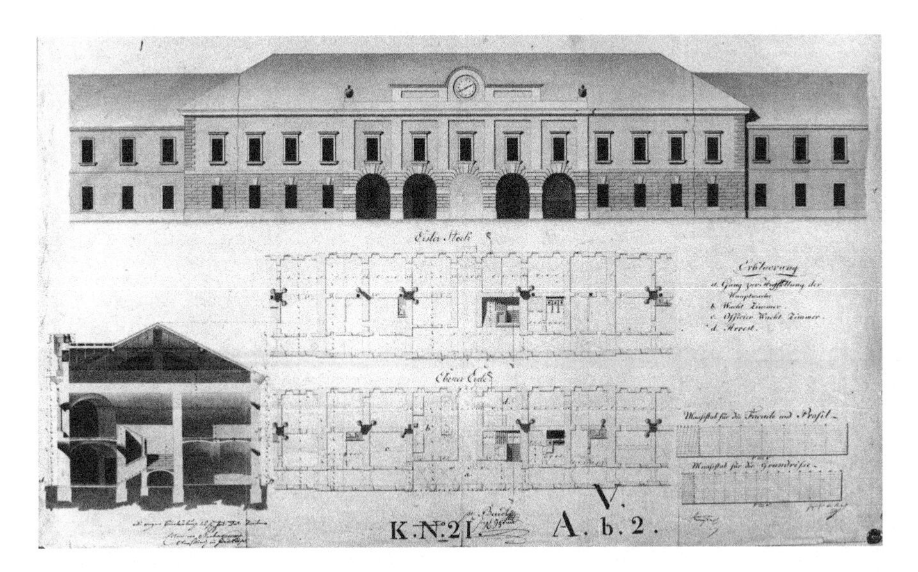

Fig. 13 The Officers' Pavilion of Peschiera del Garda, 1817 project. In the plans of ground and first floors, the stoves are very well noticed

[11]P. 91.

Fig. 14 The Officers' Pavilion, longitudinal section with cast-iron stoves and kitchens (1817 project)

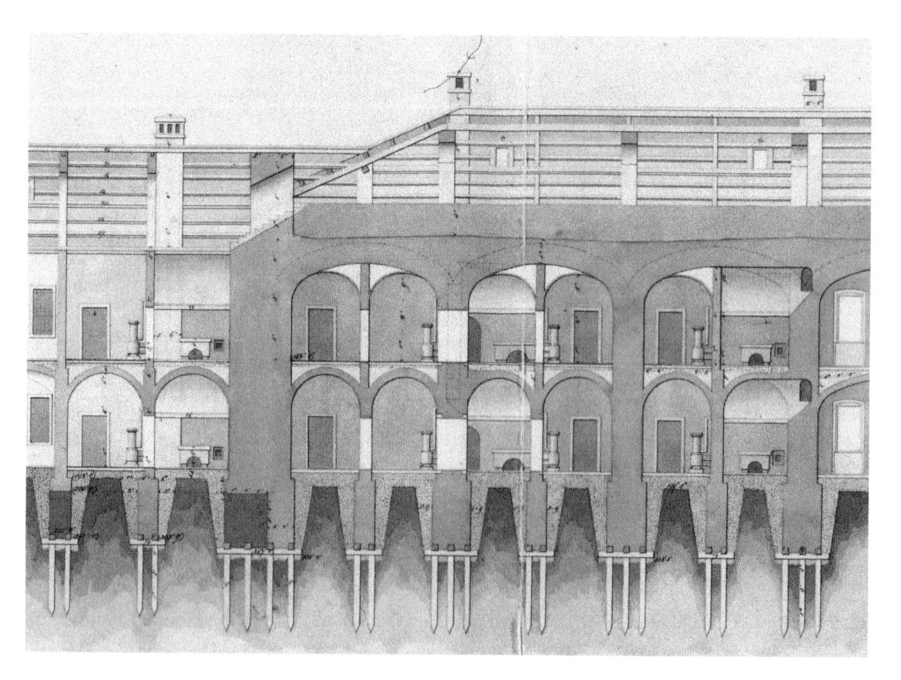

Fig. 15 The Officers' Pavilion, longitudinal section with cast-iron stoves and kitchens (1857 survey)

adorned at the top by a vase or an elegantly shaped lantern. In the same section there is also the kitchen of one of the 29 apartments, arranged in a smaller room, in which the masonry cooker was placed.

About living comfort, the placement of the stoves in the center of the lodging could generate an unpleasant feeling of cold as one approached the outer walls. This was however largely mitigated by some architectural and constructive factors: normally, the windows in the inhabited rooms were always facing south, thus allowing important solar gains in the winter season; the distribution corridor was instead located to the north and, acting as a filter with the lodgings, attenuated the effects of the colder exterior façade. Furthermore, the construction quality of these buildings should always be considered: their external walls were often made with a "bomb proof" structure, thus counting on extremely generous wall thicknesses, able to oppose an effective resistance not only to the bullets but also to the heat transfer, both in winter and in summer. The roofs had similar characteristics, as in the non-habitable attic a thick layer of earth was arranged (Fig. 15), able to cushion the effect of a possible artillery blow but also to offer effective insulation in every season. It is therefore simple to demonstrate that the living comfort inside these buildings could reach very high levels, almost unknown to the civil homes of the time but still very respectable nowadays, thanks to some significant passive measures integrated by an active and modular heating system. A system made of several high efficiency stoves, characterized in the lodgings of greater importance by an uncommon elegance, distinguishing among the other architectural and furnishing elements.

Finally, it must be said that numerous historical antecedents of the sixteenth and seventeenth centuries, related to the heating systems installed in the buildings for the military quartering, are documented in the fortresses of the Venetian period, in Verona and in Peschiera. At that time, it was customary to place, in each large room inside the barracks, a double utility fireplace: for winter heating and also for the daily preparation of food. The chimney was normally located in the middle of the back wall, opposite the windowed one, generally obtained in the thickness of the wall or by a small apse. These constructive dispositions can be observed in the relevant design of the Porta San Zeno barracks of Verona (Fig. 16a), designed by the public engineer Antonio Pasetti around 1770, and in the project of the barracks no. 1, elaborated at the beginning of the nineteenth century by the French military engineers who worked at the Direction of the Fortifications of Peschiera (Fig. 16b).

In the lake fortress,[12] however, we can still observe, even if debased by an unjustifiable carelessness, the most remarkable antecedent. Inside the mighty *Porta Verona*, or *Porta Quirina*, built in 1553 (Bozzetto 1997[13]), we meet on the sides of

[12]By affinity with the themes summarized above, we cannot omit, always in the fortress of Peschiera, the miraculous preservation, almost in a state of integrity, of the four large Hapsburg ovens of the military bakery attached to the *Provianda della Rocca*. The unique structure of the early nineteenth century is documented by Genie-Direction drawings. See Bozzetto (1997, pp. 204–205).

[13]Pp. 110–123.

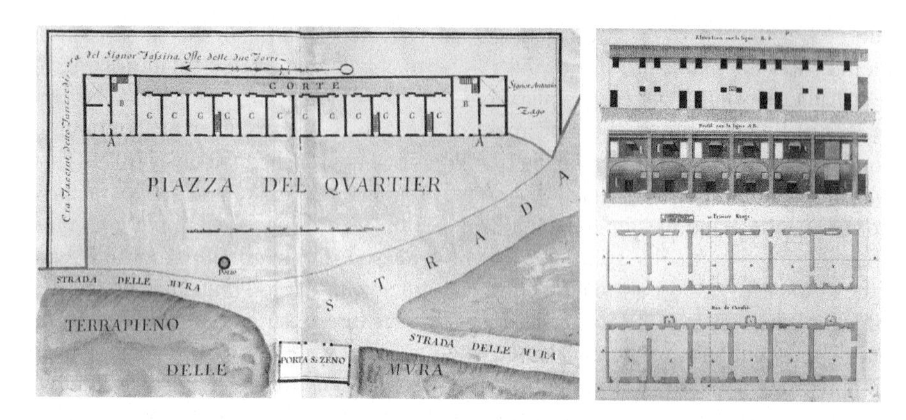

Fig. 16 Porta San Zeno barracks of Verona (1770) and barracks n. 1 of Peschiera (early XIXth century)

the hall, placed in the center and in symmetrical correspondence, two monumental chimneys with Tuscan-style pilasters, trabeation and lowered arch, nobly composed of cut-stone blocks of pink *nembro*, a beautiful veronese limestone (Figs. 17 and 18). Each fireplace is two-faced, as it is replicated, in a smaller and bare form, on the opposite side of its back wall, at the service of the guards of the fortified gate. In winter, when the heavy wooden doors were closed, the drawbridge was raised, and at the beneficial fire of the two chimneys of the entrance hall, the sentinels, awake, waited for the slow passing of the night towards the dawn. Today, without cease, the urban gate is incessantly tormented by the passage of the evil vehicular traffic. For those who can see, the central space entrance hall, covered by complex and imposing vaulted structures, still presents, in a martial form, the figurative quality of a Renaissance hall, which we can imagine illuminated by the warm and sparkling flames emitted from the marble chimneys.

5 The "Italian" Heating Systems. The Diffusion of Hot Air Heaters in the Post-unification Period

To find in Verona a further evolution of heating systems it is necessary, as anticipated, to move forward a few decades after the annexation of Veneto to the Kingdom of Italy (1866). The sudden transition from a hyper-protectionist economic system—open only to the Habsburg Empire—to a certainly more liberal one, with the opening of borders not only abroad but, above all, to other Italian regions, was the fundamental prerequisite for starting the use, also in Veneto, of new materials and new technologies in the building sector. The deep backwardness of Veneto in key sectors of the economy, such as agriculture and above all industry,

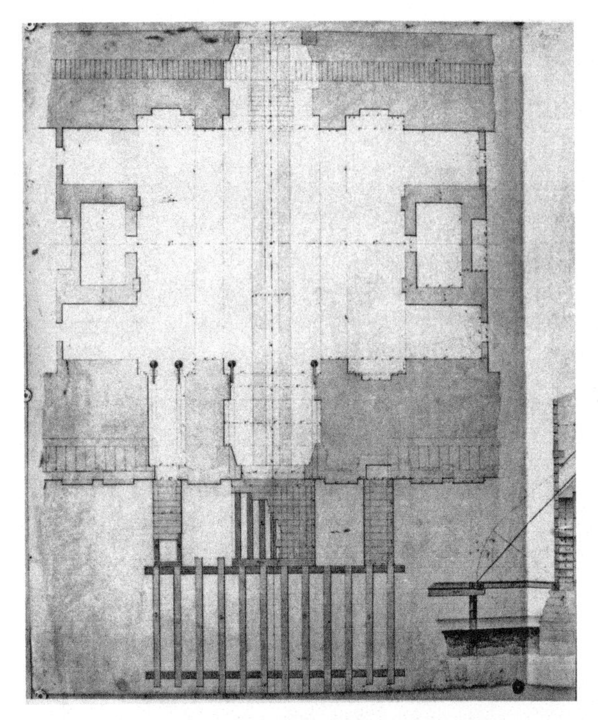

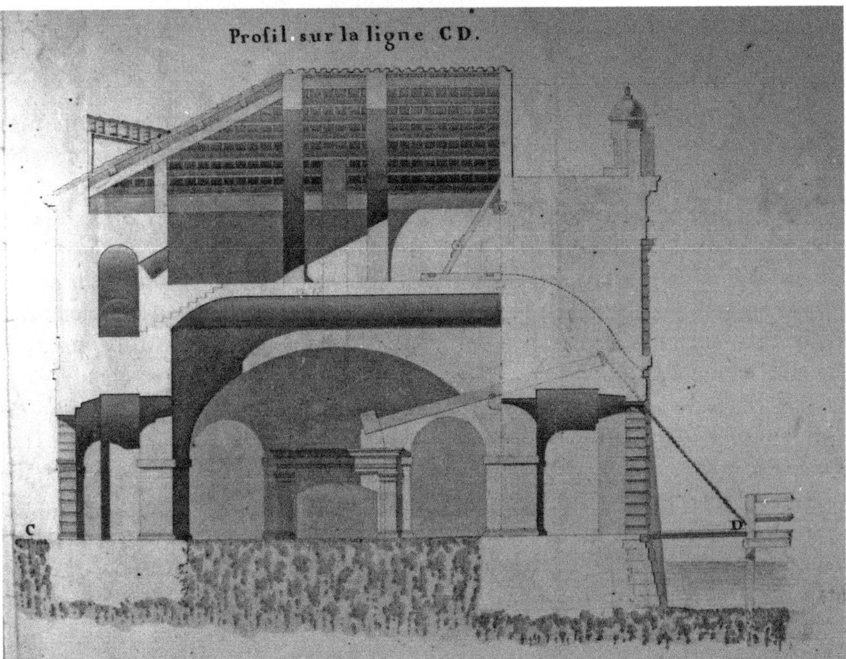

Fig. 17 Survey of Porta Verona made by the engineers of the French army in the first year of nineteenth century

Fig. 18 One of the monumental fireplace of Porta Verona, nowadays

forced the customers, at least in the first post-unitary decades, to import most of these systems from the Lombard and Piedmontese industries, if not from other European countries (Zalin 2000).

This also occurred for the installation of the new and modern centralized heating system, called *calorifero*: it was a heat generator—generally of medium or large size, fed with coal or coke—equipped with a combustion chamber separate from the air heating unit. In this way the hot air, through a complex network of ducts, was finally distributed in the various rooms and on the different floors of the building. The first three plants of this type in Verona, all built in the early post-unitary years and by 1870, were those of the Civil Hospital, the *Direzione dei Luoghi Pii* (catholic reception institutions, hospitals and orphanages) and the new Town Hall of Verona, in the important "Gran Guardia Nuova" building in Bra square, near the Arena amphitheater. In 1870, two other buildings were affected by the installation of the *calorifero* system: the Public Orphanages Institute with the annexed Maternity House and the new headquarters of the Verona Province, housed in the Prefectural Palace (formerly Palazzo Scaligero), which until the year before was occupied by the Municipality (Atti 1871[14]).

[14]Pp. 263–265.

The uses of these buildings are indicative of the benefits that the new heating system was believed to bring: firstly, to the health and comfort of fragile people, such as sick people, orphans or pregnant women; secondly, to the quality and safety of the new Municipality offices and other buildings open to the public; finally, and not less important, to the economic budget of the Municipality and other public institutions. In fact, it was well understood that central heating systems could combine fuel saving with a better and more efficient heating management and a constant air exchange in the rooms, especially inside very large and complex buildings. Therefore, this new heating systems brought with it a marked improvement in the healthiness of buildings and in the quality of life for their users.

In the 1870 report of the Provincial Council of Verona (Atti 1871[15]), regarding the installation of radiators at the orphanage, we read that this provision would have brought three types of advantages: "hygienic", because with the *calorifero* you could get an "equable temperature", a "quiet current of air in rooms easily impregnated with harmful substances", also keeping the atmosphere free from smoke; "of order and safety", avoiding the waste and the abuse of fuel compared to the indiscriminate use of many stoves; above all, avoiding "fire hazards" very probable in rooms with wooden floors and structures; finally, "economic", demonstrating the expenses incurred by the institute for heating, always higher than £ 2500 per year in the last three years, while "with the new system" the forecast was to spend less than £ 7000 for installation and only about £ 1500 per year for fuel, further saving on the cost of maintenance. In this way the expense for the plant would have been compensated in few years.

Also in 1870 the "Commission on the restoration work of the prefectural palace used by the Provincial Deputation Offices" approved a spending of £ 18,000 for the "introduction of the *calorifero* system in that building" (Atti 1871[16]). The prefectural palace complex included some of the most important and significant buildings for the history of Verona: the medieval Scaliger palaces, seat of the noble power, and the fifteenth-century Loggia of Fra Giocondo. For this reason, the commission included three important personalities such as the noble count Antonio Sparavieri and, above all, Carlo Alessandri, president of the local Art Academy,[17] and the famous architect Giacomo Franco (Conforti 1994a), one of the most important *veronese* designers of the second half of nineteenth century.

The restoration, due to the poor economic resources available, was limited to urgent repairs, postponing the major interventions on the Loggia until 1873. However, the heating system was readily installed because it was considered a source of concrete savings for the administration. With a detailed assessment, the commission established that, in addition to the costs of installation and maintenance (£ 18,000), the cost of heating would not exceed £ 3510 per year. Instead, the

[15]Pp. 263–265.

[16]Pp. 158–164.

[17]The veronese Art Academy was founded in 1764 by Gian Bettino Cignaroli. Today is a school of art and restoration recognized at the national level.

heating system in use until that time had a total of 143 stoves, for an annual use and maintenance cost of £ 5292. Distributing the expenses over twenty years, the savings with the new heating system would have been equal to £ 122 per year, in addition to all the advantages concerning to the comfort and safety of the new plant: for this reason, the Commission approved the financing for the works which were immediately executed.

6 A "Noble" Warmth: The Spread of Calorifero System in Private Buildings

Dealing with private buildings, an example of a *calorifero* system installed few years after the Unification—found through an extensive but, clearly, still partial investigation of the archival documents of Verona—concerns the noble residence of Palazzo Dionisi, in via Leoncino n. 9, a few steps from the Arena (Ferrari 1995[18]). The building was bought by Marquis Gabriele Dionisi from the Count Turchi in the first half of the eighteenth century and was subjected by the same Marquis and his heirs to numerous interventions of restoration and transformation both in the eighteenth and in the nineteenth century.

In 1873 the young brothers Gianfrancesco and Gabriele Dionisi (the latter, homonym of the ancestor who bought the palace) instructed the veronese engineer Giovan Battista Gottardi—in collaboration with the most famous brother architect Angelo Gottardi (Conforti 1994b)—of the renovation of the atrium and the entrance staircase, as well as some changes to their apartments. The intervention on the atrium is a small jewel of the post-unitary and *fin-de-siecle* architecture in Verona, studied in every detail and in many design variations by the Gottardi brothers, in which a series of veronese master craftsmen worked, including, for the pictorial decorations, the Dall'Oca Bianca brothers.

The company G.B. Monti e C. of Turin, with its branch office in Verona, was responsible for the construction of the heating system,[19] which envisaged the installation of a "*calorifero* [...] for the heating of [...] 4 rooms on the ground floor and 12 rooms on the main floor" (Fig. 19). The forecast was to heat the rooms "at a temperature that will not be less than 16 °C, with an average daily consumption from 80 to 100 kilograms of coke". In the contract with the owners, for a total amount of £ 2940, the construction of a new chimney or the adaptation of an old one was excluded, while three years of ordinary free maintenance of the plant was

[18]About the Dionisi family and the works at the building in via Leoncino, between the eighteenth and nineteenth centuries (until 1866), see pp. 305–346.

[19]ASVr, Famiglia Dionisi-Piomarta, busta 555, fasc. "Caloriferi pel riscaldamento Palazzo Dionisi in Verona. Contratto e relativi colla Ditta Monti e Comp. di Torino rappresentata dall'Ing. Ferdinando Benini". The drawings of the installation project for the new *calorifero* are dated October 19, 1873 and signed by the engineer Benini. About him see Rigoli (1994).

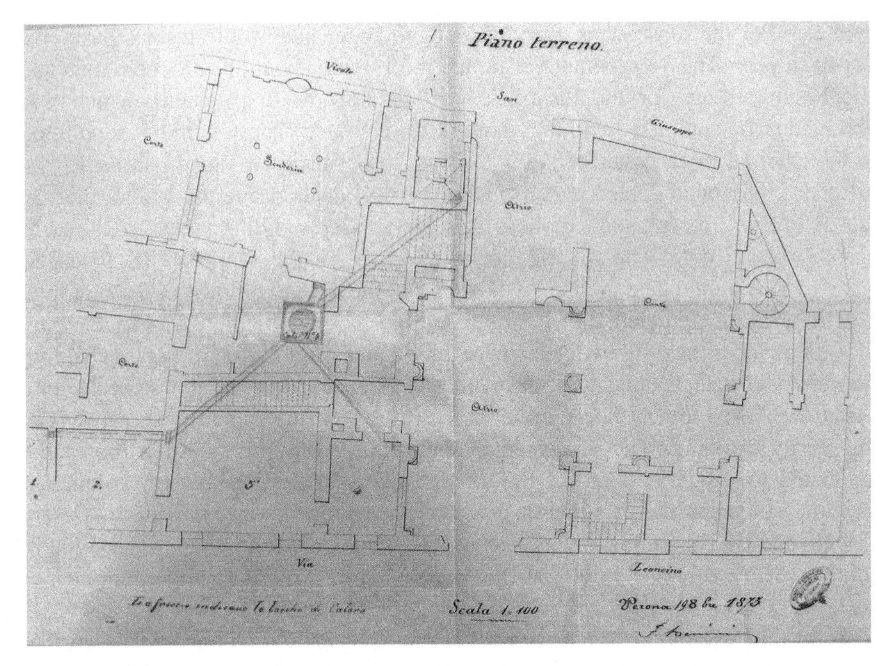

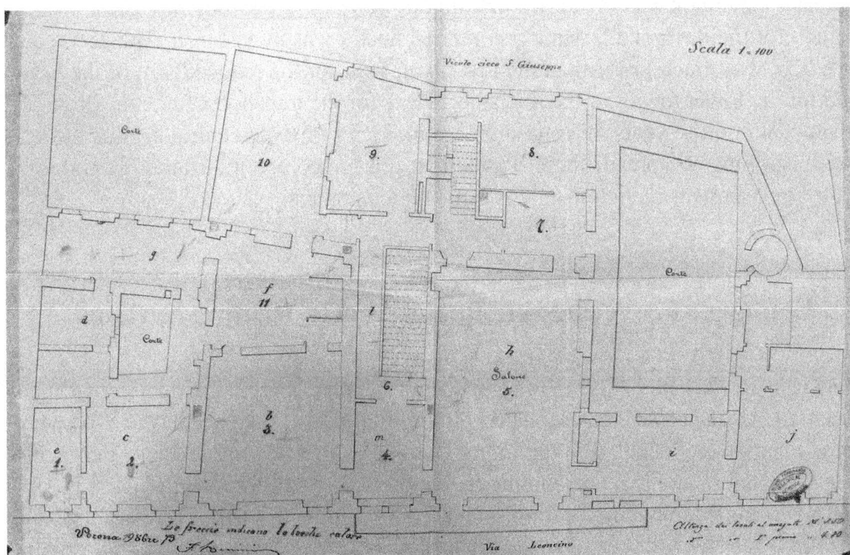

Fig. 19 Palazzo Dionisi, ground and first floor. Project for the installation of a calorifero heating system by eng. Benini, 1873

included. For the subsequent seven years the maintenance would instead have cost £ 90 per year. The very detailed estimate informs us about all the operations and supplies necessary for the installation of the new heating system, which was expected to be completed within a month. Among the masonry works—in addition to the external brick cover of the *calorifero*, to be made on site—there were "34 meters of horizontal ducts", "32 meters of vertical ducts inside the walls" and "20 meters of flue"; among the supplies, there were "9 brass tilting vents" (£ 20 each) and "3 special ones, one for the saloon and two for the staircase" (£ 30 each), clearly used to modulate the intake of hot air in the different rooms of the building.

The plant, inaugurated already at the end of 1873, was certainly appreciated by the noble clients, as only two years later, on the occasion of the important wedding between Gabriele Dionisi and the Countess Lucrezia Bembo, the same heating system was used to heat three further rooms in the "spouses apartment", always in the family building. The estimated average daily consumption of carbon coke amounted to just 15–20 kg for this second *calorifero*, for heating a volume of 443 m^3. This time the installation was realized by the Turin company of Cesare Crivelli and C. The same company, as we will see, in the following decades will in some cases provide its own hot air heating systems also to the public buildings of Verona.

From the description of the second *calorifero* of Palazzo Dionisi, we learn that the heat generator was made of "first quality refractory earth of raw color without paint", for the outer shell, while the internal heater was in cast iron. The cost, equal to £ 220, was much lower than the first plant, both for the reduced size of the heater and for the lower length of the ducts necessary for the transport of hot air. Probably, given that in those years in Verona the technology of hot-air central heating systems was beginning to spread, there was also a greater competition among producers, with a consequent reduction in costs for the customers.

7 Epilogue. From Hot Air to Hot Water Heating Systems

Between the 70s and 90s of the nineteenth century there were in Verona many installations of hot air heaters inside public buildings[20]: already in the 70s two of these plants were installed in the municipal library, housed in the ancient church of San Sebastiano, and two more in the municipal schools of Santa Eufemia.[21] In 1889 and 1890 the Municipality commissioned the famous Milanese company Edoardo

[20]From 1881 until 1935 almost all the contracts stipulated by the Municipality of Verona with suppliers of materials, services and labor are searchable and available at the Municipal Historical Archives of Verona (ASCVr), whose material is basis of this research and whose staff is thanked for the collaboration.

[21]Thanks to the maintenance contracts of these two plants present in ASCVr, after 1881, it was possible to hypothesize that their installation had taken place between 1870—as not included in the list included in the acts of the provincial council of September 12, 1870—but anyway before 1881.

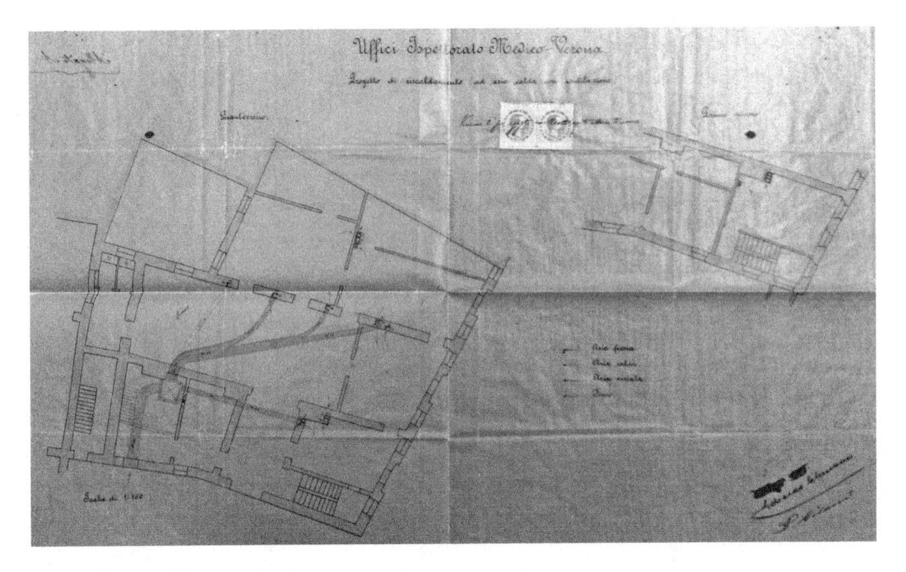

Fig. 20 Verona medical inspectorate (1889)

Lehmann to build, respectively, a *calorifero* for the new building of the medical inspectorate (Fig. 20) and other two ("Staib type *calorifero*", Sacchi 1874[22]) for a new primary school, also in Santa Eufemia[23] (where an important school complex is still housed today). In 1892, two *calorifero* systems were placed by the veronese company Eupilio De Micheli in the Verona Police Headquarters[24] (Fig. 21), inside the ancient medieval palace of Mercato Vecchio; the following year, the same company was commissioned to supply and install "three hot-air systems to heat and ventilate […] the building under construction for a female elementary school in the garden attached to Palazzo Montanari[25]", for the cost of £ 6750. These orders are particularly interesting because they concern, for the first time, two supplies made by a local company that in the following years will become the main provider of the Municipality of Verona for this type of plants.

Eupilio De Micheli was a company well known in Verona at least until the 70s, specialized in the processing of cast iron and for this reason especially engaged in infrastructure and industrial works, such as the construction of bridges and pipelines. Furthermore, it had already started a production line of painted red earthenware. In 1874 Mr. Eupilio De Micheli, member of the Academy of Agriculture of Verona, was mentioned in a speech by its President, cav. Edoardo De

[22]Pp. 628–630.

[23]ASCVr, RM.02—06/08/1889—Rep. 411 (medical inspectorate) e RM.02—28/04/1890—Rep. 706 (S. Eufemia school).

[24]ASCVr, RM.02—08/11/1892—Rep. 1496.

[25]ASCVr, RM.02—03/08/1893—Rep. 1799.

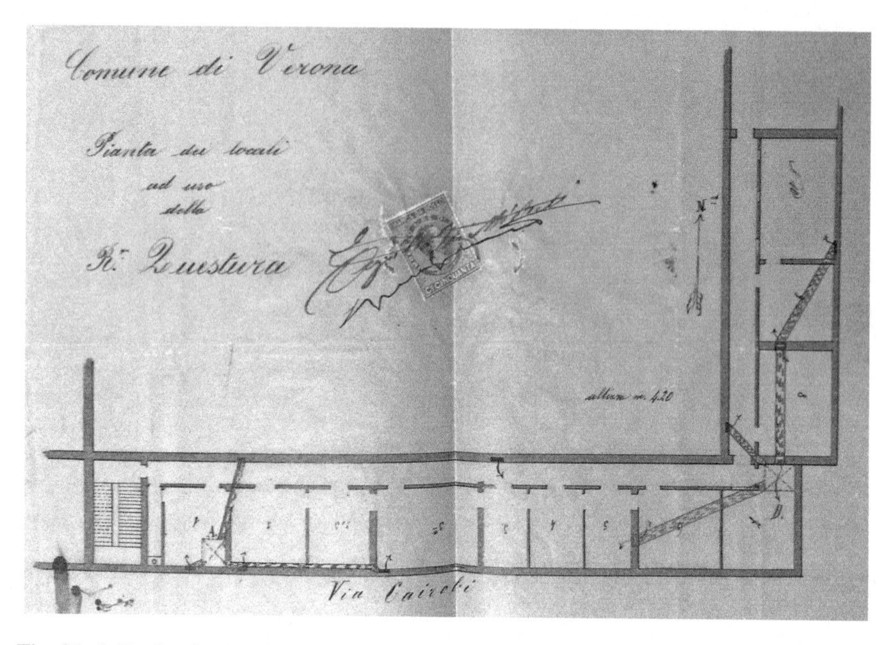

Fig. 21 Police headquarter heating system (1892). The first calorifero, called "A", took the cold air from room n. 3 and put it in the corridor and in rooms n. 1 and n. 2. The second calorifero, called "B", took the cold air from the rooms n. 8–9–10 and put it in the n. 5–6–7

Betta (Memorie 1874[26]). He stated that the De Micheli company was located in S. Vito di Bussolengo, an important town a few kilometers west of Verona, and gave work to 200–250 workers. In the following years the company started production of stoves, heaters and kitchens, enjoying great success and opening numerous offices in the main Italian cities. Mr. Eupilio De Micheli died in 1899 but the company survived for a few more years and was the most important manufacturer of heating systems in Verona for the first part of the twentieth century. In 1908 it also published an extensive catalog (De Micheli 1908—Fig. 22[27]) that illustrated both the heating systems in production and the installations carried out until that year throughout Italy. From the catalog we learn that the company had changed its location, moving to the town of San Martino Buon Albergo, just outside Verona walls. The production could count both on stoves and hot-air *calorifero*, and on more modern heating systems, such as radiators (hot water) and steam.

In the archive documents regarding the system installed in the female school of Palazzo Montanari, a very detailed drawing by De Micheli was also found, representing every minimum detail of the installed *calorifero* (Figs. 23 and 24). This was an object with "ordinary" characteristics, compared to those reported in

[26]Pp. 13–14.

[27]Thanks to arch. Michele De Mori for its report and valuable contribution.

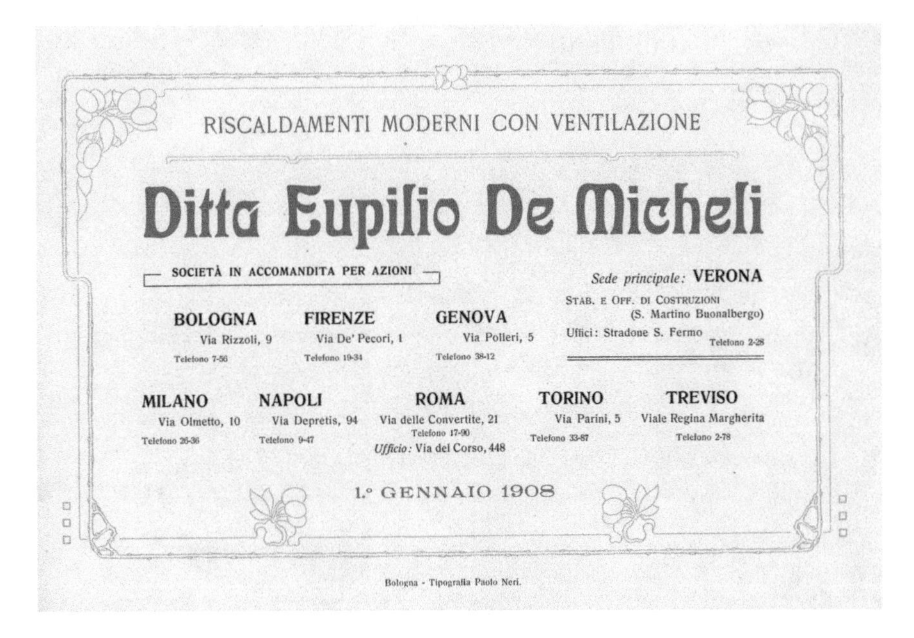

Fig. 22 First page of the De Micheli catalog, 1908

numerous publications of the time, made from a sturdy base in "common bricks" used also for the exterior coating. Immediately above the base was placed the horizontal burner made entirely of cast iron, which consisted of a combustion chamber connected to a first passage for the hot fumes. In the base there also were some water containers used to humidify the flow of hot air. The cast iron elements had a very articulated design to maximize the surface exposed to the fire and hot fumes, transferring the heat as much as possible to the clean air that surrounds the cast iron parts, all under a first shell made of "special bricks".

Above this shell there was a long coil of ducts made of special refractory brick pipes and connected, through a vertical cast iron arm, to the lower part of the burner: through these ducts an additional passage of the hot fumes took place, which allows to increase the exchange of heat with the air to be introduced into the rooms. Above the brick pipes there was a second shell, this time made of common bricks and connected to the first by a vertical cast iron pipe placed at the center of the stove. In correspondence of the upper shell, on which a thick insulating layer was placed, there were the inlets of the iron pipes which transferred the hot air into the different rooms of the building. The entire structure was surrounded by a double wall in common bricks with an insulated cavity inside.

Every room in the school had to have two adjustable outlets for the intake of hot air, the first at 0.50 m and the second at 2 m from the floor, and another one for the evacuation of the stale air, also adjustable. The temperature of air, humidified, put in the classrooms was about 60 °C in the hours between 9 a.m. and 4 p.m. The goal was to reach a temperature in the classes between 12 and 15 °C and in the corridors

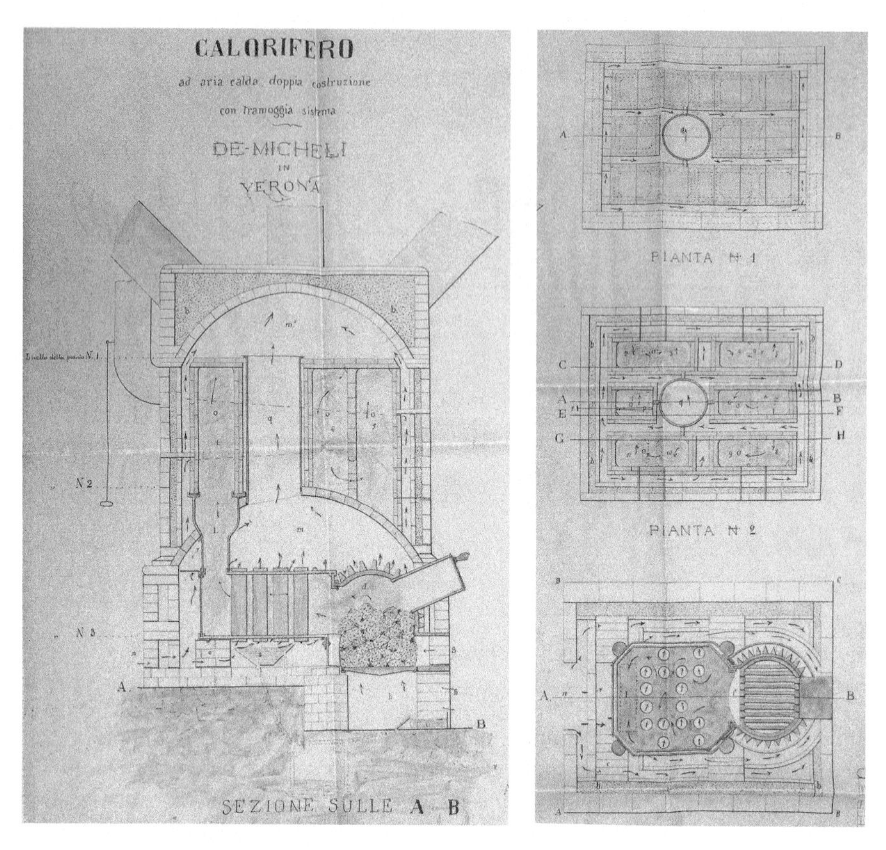

Fig. 23 Calorifero De Micheli, vertical and horizontal sections

between 10 and 12 °C, when the outdoor temperature was −4 °C. The average consumption of coke was estimated at 320 kg overall for the three heaters.

Returning to the installations of the hot-air heating systems, in 1893 it took to the Turin company Crivelli & C. the installation of a *calorifero* in the public dormitories, under construction in the former Borsalino factory[28] partially destroyed by the 1882 Adige flooding, while in 1895 the same company installed two plants in a female school.[29] Between 1894 and 1897 several maintenance works took place in the plants already built, in particular for the cleaning and repair of the air ducts, at the municipal headquarters, the library, the medical inspectorate, the schools of S. Eufemia and the female elementary school in Palazzo Montanari.

[28]ASCVr, RM.02—29/05/1893—Rep. 1718. Borsalino is the worldwide famous hat manufacturer.
[29]ASCVr, RM.02—02/02/1895—Rep. 2391.

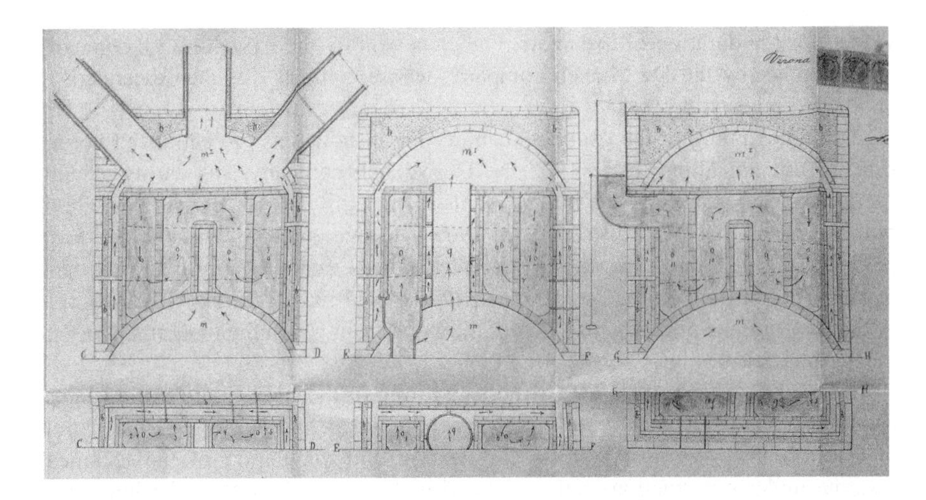

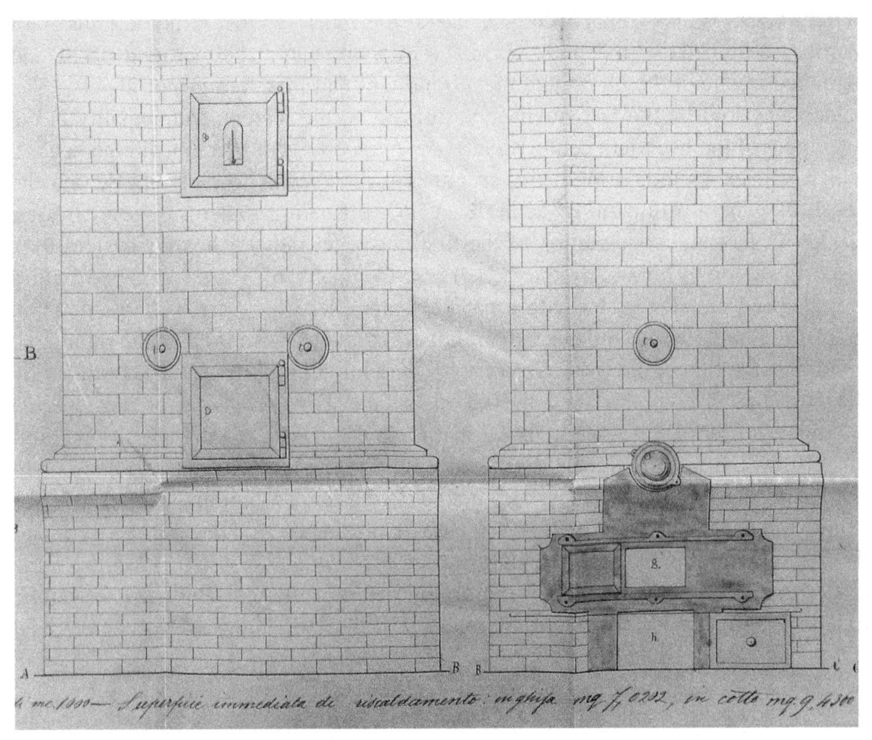

Fig. 24 Calorifero De Micheli, vertical sections and exterior facades

In 1898, given the large number of the installed *calorifero* systems in the municipal buildings (there were at least 15, in addition to those of hospitals), the Municipality of Verona decided to ban two competitions for all their maintenance works, for the duration of five years. One was won by the Crivelli & C. company and the other by the De Micheli company, whose contract was then extended by another five years in 1904.[30]

At the beginning of the twentieth century the installations of centralized hot-air systems did not diminish: in 1901 the De Micheli company was entrusted with "supplying and installing two hot-air plants for heating the San Bernardino school building in the San Zeno district[31]". In 1903 the same company was charged for a particular installation, as it was intended to realize "the heating system of a new gym under construction for the *Fiera Cavalli* (Horse Fair) of Verona[32]". All for one of the first editions of the *Fiera Cavalli*, the event from which the modern history of the Verona fair began.

In the same year (1903) De Micheli built one of the most complex and interesting installations in school buildings, which included "the supply and installation of four hot-air *calorifero* systems for heating and ventilation of the technical institute under construction in the former Bevilacqua building[33]". The interest is mainly because the work concerned one of the most significant noble palaces of Verona, designed by the great architect Michele Sanmicheli around 1530 and belonged to one of the oldest and most important patrician families of the city. The palace was sold by the last Bevilacqua duchess to the Verona Municipality at the beginning of the twentieth century and was destined to host a technical high school, which still exists today. With this installation also seems to close in Verona the phase of general diffusion of centralized hot-air heating systems. The year after (1904), in fact, the construction of the first hot-water radiator heating system took place in the civic library by the G.B. Porta Company of Turin, equipped with a "rapid-speed radiator heater with a patented propeller[34]".

In 1905 and 1906 two other radiant hot-water plants were built in the municipal buildings: the first in the *Gran Guardia Vecchia* building in Bra square, installed by the De Micheli company, and the second at the Town Hall, by of the Officina Giampietro Clerici & C. of Milan.[35] This new system replaced the hot air one built in 1870, which was previously discussed. In the following years the installations of hot air systems became rarer and linked to particular buildings: in 1907, for example, the De Micheli company created an air heater "for heating the gym and the new classrooms of the elementary school for women", while in 1908 the same company built another plant for "the drying room in the building used for public toilets" in the Campofiore area of Verona.[36] All these buildings were characterized

[30]ASCVr, RM.02—30/09/1898—Rep. 4045 and ASCVr, RM.02—01/10/1898—Rep. 4048.

[31]ASCVr, RM.02—12/11/1901—Rep. 5088.

[32]ASCVr, RM.02—31/03/1903—Rep. 5491.

[33]ASCVr, RM.02—19/09/1903—Rep. 5631.

[34]ASCVr, RM.02—19/07/1904—Rep. 5897.

[35]ASCVr, RM.02—07/10/1905—Rep. 6341 and ASCVr, RM.02—27/09/1906—Rep. 6683.

[36]ASCVr, RM.02—09/09/1907—Rep. 6993 and ASCVr, RM.02—04/11/1908—Rep. 7303.

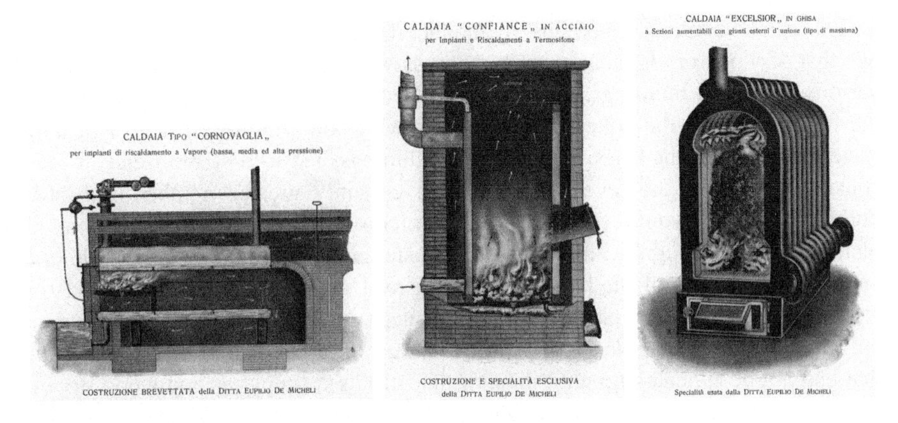

Fig. 25 Three types of stoves produced by the De Micheli company in 1908

by a discontinuous presence of people for short periods, making air heating system more convenient than the radiant one.

In 1908, the De Micheli company catalog (De Micheli 1908) included the drawings of the stoves in production (Fig. 25) and a list of the numerous installations carried out throughout Italy. Among the veronese references there are as many as 55 hot water systems, 16 hot air systems and only one steam system, built in the *Teatro Filarmonico* (Philarmonic theatre). Among the first, there are also the plants built in the civil and the military hospitals, in the offices of some important banks and companies, at the *Ristori* theater, in many private buildings and, obviously, at the offices of the De Micheli company.

In the following years no new centralized hot-air heating systems were installed, although many of them continued to be used for several years due to the investments made. In most cases they were then almost all replaced or, in some cases, flanked by those with hot water, less complex and expensive for the installation, especially in large historic buildings, cheaper and certainly simpler to manage. It is therefore possible to state that the veronese season of air heaters lasted approximately 40 years, from 1866–67 to 1907–08, and that in spite of the short duration they contributed in a fundamental way to the development of modern concepts of comfort and safety related to the heating systems inside buildings.

A final note regards the consumption of fuel, that is coke. Thanks to the contracts of the suppliers deposited in the municipal historical archive, it was possible to trace the amount of coke purchased by the Municipality of Verona for all the years between 1884 and 1912. Consumption grew rapidly between 1884 and 1890, passing from 50 to 200 tons per year, and then remained essentially stable until 1906. From 1906 to 1912 there was instead a further increase, going from 200 to 400 tons per year. As can be seen by comparing the consumption trend with the installation of new heating systems in the municipal buildings, the first substantial increase in consumption in the 80s did not correspond to a second in the 90s, although in the latter period there were several new installations.

This can be explained, perhaps, by the fact that the use of coke supplanted in the 80s that of wood or charcoal even in many stove systems, certainly still present in various municipal buildings. If this were indeed demonstrated, it would mean that the replacement of the coke-powered stoves with new centralized hot-air heating systems, added to the ones made in new buildings, led to a stable trend in total consumption. This, however, faced with a certainly greater number of heated buildings, which would confirm the lower consumption of centralized hot-air plants. Instead, the significant increase in consumption occurred between 1906 (the year of construction of the hot-water heating system in the Town Hall) and 1912 would seem due to the rapid spread of hot water systems, which in a few years not only replaced the old *calorifero* plants, but were also adopted in many buildings that had not been heated up until then, or that still used fireplaces and wood stoves. The twentieth century, with its exponential increase in energy consumption, had begun also in Verona…

References

Atti (1871) Atti del Consiglio Provinciale di Verona, Anno IV, 1870. Verona, pp 263–265

Berengo M (2009) La società veneta alla fine del Settecento. Roma

Bernardello A (1996) La prima ferrovia tra Venezia e Milano. Storia della Imperial-Regia privilegiata strada ferrata ferdinandea lombardo-veneta. 1835–1852. Venezia

Bertoncelli G (1838) Lignite della Purga di Bolca, nella provincia di Verona. In: Annali universali di statistica, economia pubblica, storia, viaggi e commercio, Milano, pp 216–222

Bevilacqua Lazise I (1816) Dei combustibili fossili esistenti nella provincia Veronese e d'alcuni altri loro contigui nella provincia vicentina e nel Tirolo non che del loro uso come succedanei dei combustibili vegetabili. Memoria mineralogico-economica. Verona

Bozzetto LV (1997) Peschiera storia della città fortificata. Verona, pp 185–260

Cattaneo C, Milani G (2001) Ferdinandea. Scritti sulla ferrovia da Venezia a Milano. Firenze

Conforti G (1994a) Giacomo Franco (eds: Brugnoli P, Sandrini A), pp 441–446

Conforti G (1994b) Angelo Gottardi (eds: Brugnoli P, Sandrini A), pp 454–458

Da Lisca A (1841) Sul combustibile fossile di Purga di Bolca, e sopra ciò che difficulta il suo traffico. In: Annali universali di statistica, economia pubblica, storia, viaggi e commercio. Milano, pp 88–93

De Micheli DE (1908) Riscaldamenti moderni con ventilazione. Bologna

Ferrari ML (1995) Nobili di provincia al tramonto dell'antico regime. I marchesi Dionisi di Verona (1719–1866). Verona, pp 305–346

Ferrari ML (2012) Quies inquieta. Agricoltura e industria in una piazzaforte dell'Impero asburgico. Milano

Landi AG (2010) Impianti di illuminazione a gas nel XIX secolo. Genesi e sviluppo delle reti urbane nel Regno Lombardo-Veneto. PhD thesis, Politecnico di Milano

Memorie (1874) Memorie dell'Accademia d'Agricoltura Arti e Commercio di Verona, vol LII, Serie II, fasc. I. Verona

Perbellini G, Bozzetto LV (1990) Verona: la piazzaforte ottocentesca nella cultura europea. Verona

Preto P (ed) (2000) Il Veneto austriaco 1814–1866. Padova

Rigoli P (1994) Benini Ferdinando (eds: Brugnoli P, Sandrini A), pp 454–458

Sacchi A (1874) Architettura Pratica—Le abitazioni: alberghi, case operaie, fabbriche rurali, case civili, palazzi e ville. Milano, pp 628–630

Scalva G (2010) L'invenzione di Benjamin Franklin e l'attività manifatturiera di Castellamonte. In: I quaderni di Terramia—8. Torino, pp 44–50

Scopoli G (1841) Della ricerca del carbon fossile. In: Memorie dell'Accademia d'Agricoltura Commercio e d'Arti di Verona, vol XIX. Verona, pp 33–47

Selvafolta O (1994) Verona Ottocento: i luoghi e le architetture dell'industria (eds: Brugnoli P, Sandrini A), pp 195–261

Tonetti E (1997) Governo austriaco e notabili sudditi. Congregazioni e municipi nel Veneto della Restaurazione (1816–1848). Venezia

Zalin G (2000) Il tempo di Giovanni Calabria (1873–1954): economia e società a Verona e provincia tra Ottocento e Novecento. In: Bollettino dell'archivio per la storia del movimento sociale cattolico in Italia, vol 35, no 2, Milano, pp 131–143

The Alte Pinakothek

Melanie Bauernfeind

Abstract The Alte Pinakothek, opened in 1836 as one of the early free-standing museum buildings in Europe, has a rich and diversified history. The innovative aspect of this contribution is the reconstruction of the building history with focus on its technicals aspects. This turns the Alte Pinakothek also into a vivid example for the influence of developments in science and engineering on the "museum world" with its specification for the preservation of works of art—and vice versa. More than 175 years history of past approaches with specific technical solutions demonstrate the complex interactions between the architectural concept, the building envelope and the site specific conditions. These considerations are complemented by hygro-thermal modelling and light simulations which for the first time document the development of historic conditions influencing the preservation of works of art. By analysing the simulations of historic conditions, the efficacy of simple technical interventions of the past is revealed and contributes to future perspectives in museum architecture.

The Alte Pinakothek is one of the early free-standing museum buildings in Europe. It opened in 1836 and over the decades had a rich and diversified history. The presented innovative attempt to reconstruct these 175 years of this building and its related technical history should contribute to future perspectives in museum architecture. The past of the Alte Pinakothek and in particular its different concepts of climate control demonstrate how different criteria of preventive conservation interact. Despite these aspects, it is a descriptive example for the influence of developments in science and engineering on the "museum world" with its specification for the preservation of works of art—and vice versa.

Thereby, the architectural concept, the building envelope and the building materials gave distinction to the indoor climate conditions and the technical effort to achieve them. These correlations and the specifications for human comfort in the past often resulted in over-dimensioned HVAC systems for museums, which were

M. Bauernfeind (✉)
Bayerische Staatsgemaldesammlungen, Doerner Institut, Munich, Germany
e-mail: Melanie.Bauernfeind@doernerinstitut.pinakothek.de

C. Manfredi (ed.), *Addressing the Climate in Modern Age's Construction History*,
https://doi.org/10.1007/978-3-030-04465-7_3

cost and energy intensive. Today, this turned out to be a burden for the preservation of our cultural heritage. However, an increased concern to conserve natural resources combined with the financial limitations faced by most museums and historic houses has generated discussion about whether it is still suitable to run museums with this effort and high technical invent. In addition, the high energy consumption of air-conditioning systems is a major financial drain, exacerbated by their operation and maintenance costs.

This review of past approaches and available technical solutions for the Alte Pinakothek demonstrates how important it is to understand the complex interactions between the architectural concept, the building envelope and the site specific conditions. These considerations are complemented with hygro-thermal and light simulations to reconstruct the different phases of the buildings history. The varying historic indoor climate conditions and the necessary energy consumptions are modelled with input from archival sources such as building plans, blueprints, correspondence, submissions, invoices, etc. By analysing the simulations of past indoor climate conditions, the efficacy of the simple technical interventions of the past is revealed, enabling an insight into the architectural expertise of those involved with the museum over the decades.

1 Museum Climate and Technical Solutions

Throughout human history attempts have been made to control indoor climate conditions. Whereas the early focus was to ensure human survival, as technologies developed the demand for comfort increased. A comparable trajectory can be traced for climate control in museums to be achieved by heating, ventilation and air-conditioning (HVAC) systems. Initially, recommended levels for relative humidity (rH) and temperature were based on practical observations. Later, an increased awareness of specific material characteristics and a growing understanding of physical and chemical deterioration processes influenced climate recommendations.

Climate set points in conservation are defined values for rH and temperature to ensure stable conditions for the preservation of artworks. These often contain hygroscopic materials and combinations that react to changes in rH by swelling or shrinking. This phenomenon has been studied over the last 80 years and has resulted in the generally accepted understanding that rH values below 30% cause embrittlement and shrinkage, whereas values above 70% cause swelling and mould growth. Specification of absolute values within this range seems to be more a question of belief or of subjective interpretation of real-life observations rather than being founded upon scientifically proven findings. Today's set points are supposed to originate during World War II when objects were evacuated into quarries. This happened in many European countries involved in these conflicts, including

England,[1] Austria (Frodl-Kraft 1997) and Germany (Schawe 1994, 2010). In most cases, like for example in England simple heating systems were used to keep the environmental conditions constant at about 58% rH and 17 °C. The fact that no damage occurred was ascribed to the absolute values rather than the extreme stability of the climate. Interestingly, one of the earliest documented set point values was established far earlier, that is to say at the end of the nineteenth century. It is particularly noticeable that human comfort criteria rank lower than achieving the correct conditions to preserve works of art. Moreover, set points in the past often mirror technical feasibilities. The more powerful the technical installations became, the stricter were the set point values and their ranges of acceptable deviation.

Within todays' museum world, strict requirements are commonly defined for entire collections, irrespective of site, building envelope or local museum architecture. Thereby, the relationship between museum architecture and indoor climate conditions is generally ignored. Instead, complex technical systems are installed to compensate inadequate building physics. In contrast, in early museum history, indoor climate conditions were chiefly influenced by the massive building structure. Moisture control meant protection from driving rain and leaky roofs. If there was any heating, this was provided by the combustion of coal or wood. Natural ventilation arose by the airflow through ducts and chimneys or the buildings leakiness. Besides, window openings had to be large enough to ensure sufficient day lighting.

At the end of the 19th century, issues like ventilation and filtration gained importance since central heating was introduced also for domestic homes and since the first gas lighting systems released harmful emissions. Due to the high levels of pollution outdoors, as a consequence of the growing industrialisation natural ventilation was counterproductive. The birth of air conditioning technology and the initial point of all further developments either in museum architecture was the ambition of hygienists and engineers to eliminate this deficit with mechanical ventilation and air purification. In consequence, mechanical ventilation became an urgent challenge. All these factors, ventilation, artificial light and the quality of the building envelope, affected the indoor climate in the past as they do today.

The investigation of the technical building history of the Alte Pinakothek has quarried interesting information on historic strategies of climate control and lighting of the gallery. This information and data served as input for the development of virtual models to simulate the indoor climate conditions as well as the illumination situation over the decades.

[1]http://www.nationalgallery.org.uk/paintings/history/the-gallery-in-wartime/ (accessed 19 April 2013).

2 The Alte Pinakothek and Its Technical Building History

The Alte Pinakothek was one of Europes first public picture galleries. It was built by the architect Leo von Klenze between 1827 and 1836 for King Ludwig I in collaboration with its first director Georg von Dillis. In fact, it was Dillis's 14-point catalogue of requirements, the so-called 'Prememoria' (Eibl 2011) that had greatest influence on Klenze's architectural design. It is worth mentioning that this building strongly influenced subsequent generations of architects and for many decades became a prototype of museum architecture.

The computer simulations necessitated to simplify the technical history which was therefore classified into six phases of different conditions and characteristics (Table 1). In the context of this book, the periods between the opening of the gallery until the heavy destruction of the museum during World War II are described in detail. Further information on everything that happened afterwards as well as more comprehensive descriptions of the performed hygro-thermal and light simulations can be found in the authors Ph.D. thesis (Bauernfeind 2016).

3 From 1836 to 1841: Phase 1 or the Question of Heating

Like mentioned above, the gallery opened after a planning and construction period of about ten years in 1836. At that time, the building was located outside Munich's city wall, which permitted Klenze to place the longitudinal side of the museum in a north-south orientation. This concept of a fully-detached building allowed the galleries to be lit by daylight, minimised the risk of fire, avoided the noise of carriages and also reduced the ingress of dust and dirt.

In total, the building (Fig. 1) is 150 m long with a width of about 50 m. Both ends have wider front sections which housed the entrance hall, with a huge staircase at the east end and functional rooms at the west. The building has three main levels: the cellar with air-heating ovens placed in so-called 'Klimakammern' (climate chambers) (Fig. 2), the ground floor, where the print and drawing cabinet, a collection of vases, a vestibule, and storage rooms were located, and the upper floor with eight galleries and 23 cabinets. Above this is an attic crowned by 11 light

Table 1 Periods of the building history of the Alte Pinakothek from phase 1–6

Period	Short description
1836–1841	Original building with air-heating planned by Klenze
1841–1891	Unheated building after deactivation of the air-heating
1891–1952	Low-pressure steam heating system for heating and humidification
1952/57–1994	Reconstruction by Döllgast with HVAC for heating and humidification
1997/98–2008	Overall refurbishment with installation of a full HVAC system
2008–today	Energy-efficient retrofitting of a single gallery room

Fig. 1 Historical drawing of the Alte Pinakothek from 1900

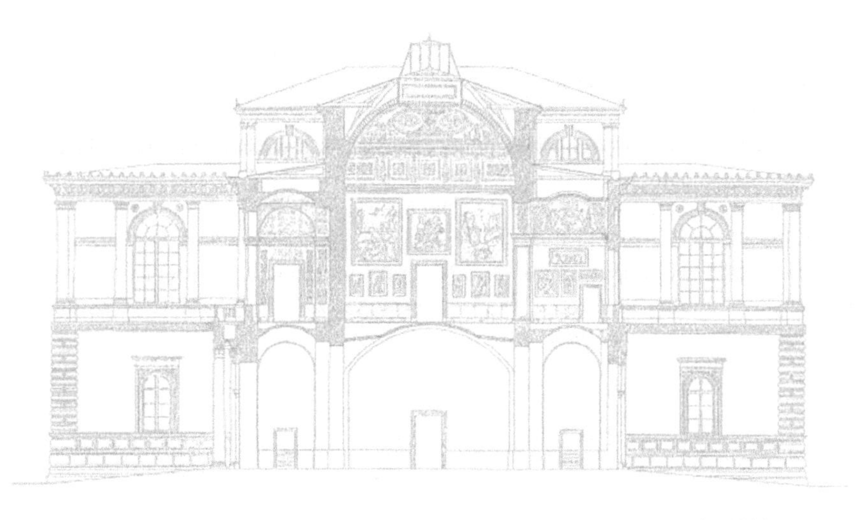

Fig. 2 Cross-section of the Alte Pinakothek, architectural drawing by Klenze in 1831

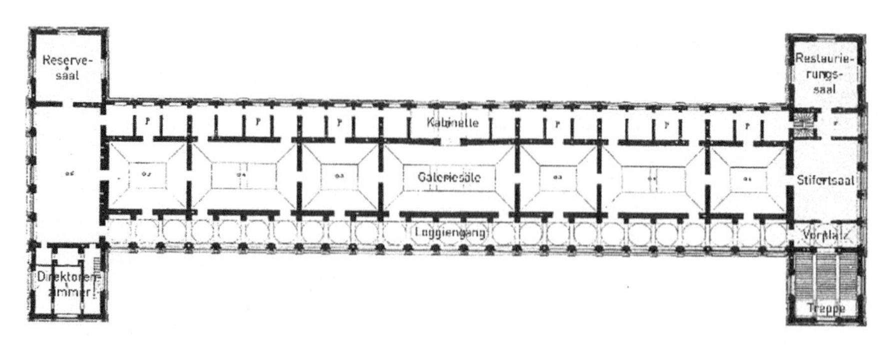

Fig. 3 Ground plan of the Alte Pinakothek. Architectural concept with tripartition of the first floor from 1836 until the heavy destruction during World War II

lanterns in the style of greenhouses, which were constructed by Klenze in response to the problem of daylight illumination under challenging weather conditions. The building envelope is heavy brick masonry faced with yellowish sandstone. The architectural ornamentation consists of green 'Regensburger' sandstone. The roof is covered with sheet copper.

Figure 3 illustrates the tripartition of the Alte Pinakothek: the major gallery rooms with skylights in the middle of the building, the side-lit cabinets in the north and the abundantly decorated loggia corridor as a passageway and climate buffer in the south. The interior design also reflects the different illumination and uses of the rooms. Whereas the ceilings of the smaller cabinets were flat, those of the large galleries had a special construction to bear the light lanterns. The loggia consisted of 25 pendentive cupolas. All ceilings were richly decorated with partly gilded plasterwork. Originally, there were terrazzo floors in the whole building, except for the vestibule that had tiles and the grand staircases that were made of marble.

Klenze argued that a heating system was required for the preservation of the paintings and for visitors' comfort (von Klenze 1930). Being aware of the fire risk caused by any heating system, he planned 14 masonry niches in the cellar in which he placed the wood-fired air-heating ovens. Fresh air, as well as heated and exhaust air, were channelled through brick ducts placed in the walls of the building which were about one metre thick. In the gallery rooms themselves, warm air was supplied through vents directly below or next to the paintings. The exhaust air was extracted about 15–20 cm above the floor. The smoke ducts were also routed through the massive walls to chimneys on the roof. Natural ventilation was provided by leakage through the building envelope, air exchange between the different rooms and by opening the windows manually.

3.1 *Air-Heating as an Alternative to Fired Stove Heating*

During the time when Klenze was concerned with the question of heating the museum, air-heating has become common practice for public buildings and was state of the art (Peclèt 1860, 1986; Meissner 1827; Perkins 1841). Meanwhile the principle of steam-heating, invented by James Watt in 1769 and used for heating his fabric and domestic rooms in Birmingham, was still almost unknown in Germany in the 1830s. Not until 1843, seven years after the opening of the Pinakothek, the first steam-heating in Germany was introduced in Sigmaringen castle (Heyl 2012). This facility was constructed by Johannes Haag, who imported the hot-water ductwork and steam-heating system—invented by Perkins in England (Davies and Ryder 1841). In the middle of the 19th century, Haag began his work for the Munich nobility. He installed, for example, a steam-heating system in king Ludwig I railway waggon (Haag 1873) and between 1860 and 1865 a central-heating in the greenhouses of the royal botanical garden in Munich (Voit 1867). This leads over to the Alte Pinakothek because Haag's company was absorbed by the Sulzer AG, which at the end of the 19th century constructed the low-pressure steam-heating system for the Alte Pinakothek (Heyl 2012).

But due to the requirements and the available technical solutions at the beginning of the century, Klenze in fact had no alternative to the air-heating. His main concerns were the absence of smoke and soot, the prevention of smell, the reduction of fire danger as well as the economic utilisation of combustible materials (Anonym 1836). The effectiveness of the system increased if a building had massive walls with appropriate heat storage capacity. But following Peclét (1986) there were also other advantages like aesthetic considerations, safety and economy of space. Hence, the air-heating system of the Alte Pinakothek was a result of practical aspects as well as aesthetic and protective specifications which had to be met by a heating systems used in a museum building. Additionally, any stove heating would have failed to heat the enormous air spaces of the Pinakothek's large galleries.

The principle of an air-heating system is to place large fired furnaces in the cellar. These heat the air which in most cases reaches the rooms through ducts in the walls by natural draft. According to the way how the air circulates it is distinguished between circulating air-heating (the cooled exhaust air is returned to the heating chamber) and ventilation-heating (fresh air is added to the system) (Meier 2011). Following the invention of Paul Traugott Meissner, the system for the Alte Pinakothek combined circulating air-heating and ventilation. Therefore, the cooled exhaust air from the gallery rooms was mixed with a certain amount of fresh air before it was reheated again (Meissner 1827).

3.2 Disadvantages of the Air-Heating System

From the beginning, those air-heating systems were judged sceptical because the amount of necessary ducts resulted in high construction costs for heavy wall thicknesses. Also statical reasons interdicted the retrofit of existing buildings. In practice, the combination of heating and ventilation was another drawback because they complicated any operation and therefore hindered a purposeful temperature control. Besides, the air-heating caused a permanent draught of very dry air that spread dust and acoustic emissions via the ducting through the whole building (Meier 2011).

Only after few years in operation the described problems of the air-heating also occurred in the Alte Pinakothek. Indeed, the system avoided temperatures below zero degrees during wintertime but it entailed an extremely low relative humidity which was harmful to the works of art. The enourmeous dimensions of the gallery rooms in combination with the lack of mechanical ventilation and the small transversal section of the ducts necessitated supply air with temperatures above 60 °C and a high flow velocity, if an adequate heating of the rooms should be permitted. This, in conclusion lowered the humidity levels. The position of the air inlets nearby or marginal underneath the paintings directly exposed them to the hot and dry air. Another harmful aspect were temperature gradients arising over the immense hight of the rooms. They caused the formation of microclimates and, in consequence, damages to the paintings.[2]

3.3 Indoor Climate Conditions and Air-Heating

The results of the hygro-thermal simulations for the Alte Pinakothek prove the described problems of low rH while running the air-heating system. In practice, during the heating period from October to April, rH in the largest gallery (Rubensgallery) dropped to values below 20% (Fig. 4)—a condition extremely dangerous and the reason for the recorded damages. In contrast, the air-heating operated in a way that temperature could be kept between eight and twelve degrees. Regrettably, within a few years of opening, the air-heating system was acknowledged to be a failure. In 1841, the system was switched off due to severe damage observed on many of the gallery's paintings. In particular, large temperature fluctuations had caused even larger fluctuations in rH, and there were dust and dirt accumulations on painting surfaces.

[2]BSTGS registry file 16/3 no. 314, newspaper article Zweites Morgenblatt, Nr. 31, 31. January 1892.

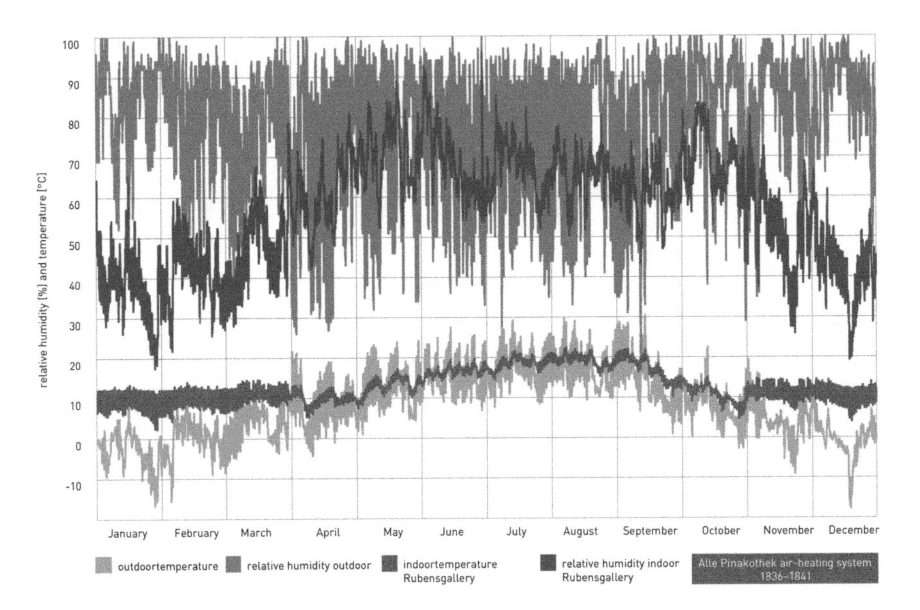

Fig. 4 Relative humidity and temperature over a period of one year when running the air-heating system (1836–1841)

4 From 1841 to 1891: Phase 2 Without Heating

Eventually, for preservation reasons, the galleries were not heated from 1841 to 1891. Furthermore, maintenance and housekeeping was not carried out during this period due to a lack of financial support from the Bavarian state. Fatal enough, this culminated in decreasing attendance and the Alte Pinakothek vanished gradually from public concern. Within several years, the formerly splendid and famous picture gallery has turned into a desolate and ruined image of herself.

4.1 No Heating as Alternative to Air-Heating

Abandoning the heating led to room temperatures as low as −5 °C during winter, which resulted in blooming on painting surfaces and mould growth. Furthermore, the building and its installations had also suffered damage.

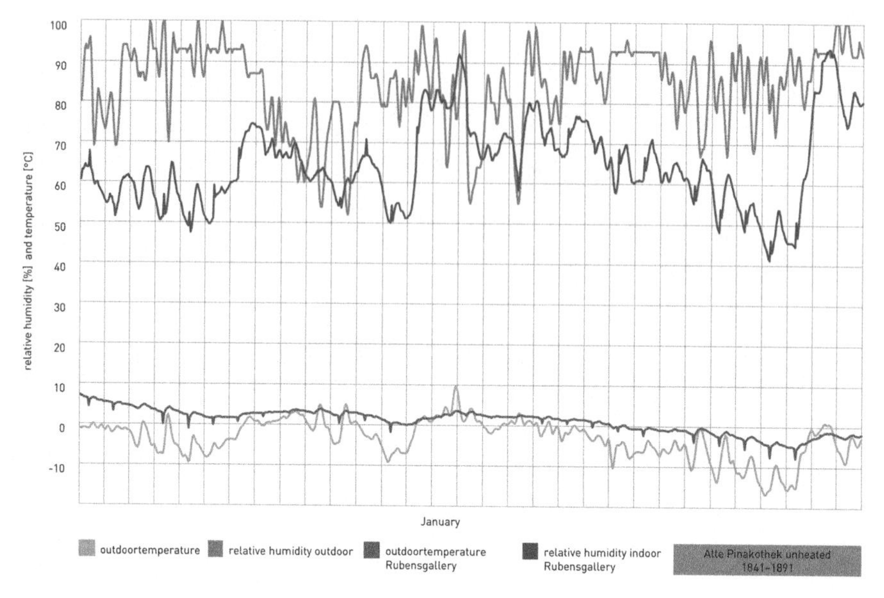

Fig. 5 Relative humidity and temperature during January for the period without any climate control (1841–1891)

4.2 Passive Indoor Air Conditions: Influence of the Building Envelope

Analysing the simulated data (Fig. 5) shows striking stable temperatures, which can be traced back to the buffering effect of the building envelope. Both, heat storage capacity of the massive brick walls and the moisture buffering effect of the wooden panelling reduced the fluctuations. Nevertheless, the indoor climate conditions with temperatures of averaged 5 °C and relative humidities ranging from 60 to 90% were harmful. Besides the problem of condensation on the glazed light lanterns, temperatures ranging from minus ten to plus 30 °C were challenging. The differing expansion coefficients of the construction materials entailed permanent leakiness, which was serious enough to promote the ingress of rain. Solely the lay-lights, as second glazing layer, hindered water drops to trickle into the galleries.

4.3 Development of Air Conditioning Technology in the 19th Century: Systems and Improvements for Museums

Until the end of the 19th century and over a period of about 40 years the conditions in the Alte Pinakothek remained the same. However, in the middle of the century,

Max von Pettenkofer had proven that in particular humidity fluctuations caused severe damaged to the painting surfaces. For this reason, a special commission was mandated to develop a new climate control strategy for the museum. Such a system should ensure rH to be kept far away from any evaporation or condensation.[3] Up to now, the evacuations during World War II were presumed to be the hour of birth of today's set points. But the sources proof that they were pronounced 100 years before in Munich.

The second half of the 19th century is affected by numerous developments in HVAC technology and control of indoor conditions. The progress expedited by the industrialisation yielded new materials and manufacturing methods, which allowed for alternative climate control strategies. Striving for ideal museum specific solutions and achieving indoor climate conditions that enhances the preservation of artworks, some of them were applied in the new museum buildings of that time. Hence, it is worth to throw a glance at the strategies established in other museums in Germany. They exemplify the range of preventive measures carried out to better preserve cultural heritage. Moreover, the experiences with these systems were decisive for the choice of the low-pressure steam-heating, which finally was implemented in Munich's Pinakothek.

4.4 Old National Gallery in Berlin: Hot-Water Air-Heating

In 1876, Berlin's Old National Gallery, designed by Friedrich August Stüler opened, after a construction period of ten years. The museum was equipped with a combination of hot-water heating and air-heating (Maaz 2001). The system consisted of hot-water coils placed in cased niches in which the air was warmed before it was transported to the rooms via ducts. Room temperatures could be kept at about 16 °C.[4] The cooled air was again channeled to the niches by grill covered ducts placed in the floors.

But defective seals in the hot-water conduit caused recurring leakages, which were one of the main draw backs of this hot-water heating. Subsequently, steam exhausted even in the galleries close to the works of art. Another problem was the inaccessibility of the cased coils, which made any cleaning almost impossible. Hence, the accumulated dust burnt and this caused an unpleasant smell. Besides, the dust transported by the warm air deposited on colder objects and entailed their discolouration. Like in most museums with an air-heating system, also in the Old National Gallery low rH caused severe damages to the paintings.[5]

---

[3]See footnote 2.

[4]Amtlicher Bericht der Direction der k. Nationalgallerie vom 19. April 1877, zitiert nach Protokoll der 1ten Sitzung des Reise-Ausschusses der Commission für Beheizung der k. Pinakothek vom 27. Dezember 1880.

[5]See footnote 4.

4.5 Royal Museum in Dresden: Low-Pressure Warm-Water Heating

In 1854, a low-pressure warm-water heating was installed at the Royal Museum in Dresden. The special about this system was, that the rooms of the second floor were solely warmed by openings in the floor. These allowed the warm air from the first floor to stream in.[6] This system seemed to be a success. According to Heinrich Dehn-Rotfelser, the same principle was applied to the National Gallery London, the South Kensington Museum London and the Royal Academy London (von Dehn-Rotfelser 1879).

4.6 Picture Gallery in Kassel: Conservation Compliant Heating

The architect Dehn-Rotfelser was also responsible for the Kasseler museum opened in 1877. Both, the inner partition and the lighting concept exhibit analogies to the Alte Pinakothek. But because the air-heating has turned out to be a failure in Munich, a warm-water heating was installed with which the gallery rooms were heated to maximum 10 °C and minimum 5 °C (von Dehn-Rotfelser 1879).

Again, boilers were located in the cellar. The heating pipes were placed in ducts in the floors, which were covered with iron grills. In order to avoid high temperatures close to the works of art, nearby the paintings these grills in turn were covered with plates (von Dehn-Rotfelser 1879). Those measures as well as the quite low set-point temperatures must are clearly actions to enhance the conditions for the sake of artworks.

5 From 1891 to 1945: Phase 3 and the First Attempt to Control rH

Finally, the poor condition of the Alte Pinakothek led to a major renovation in the late nineteenth century. The gallery floors were covered with oak parquet to replace the former terrazzo floors. The walls were given a new textile covering, and the copper cover of the light lanterns was replaced by depolished glass. Additionally, a low-pressure steam heating system was employed not only to heat the galleries but also to humidify them to some extent. The conditions for the paintings had priority over human comfort. The installed heating system consisted of four wrought-iron

[6]Protokoll der 4ten Sitzung des Reise-Ausschusses der Commission für Beheizung der k. Pinakothek vom 31. Dezember 1880.

low-pressure steam boilers located in the cellar. These supplied steam to the 110 ribbed radiators in the building. In the cabinets, the radiators were placed in the window recesses. In the galleries, the radiators were integrated into 13 so-called 'divans', which were large pieces of upholstered furniture in the middle of the rooms. Inside these 'divans', water basins were placed above the heating pipes to humidify the air by evaporation. During the heating period, up to 250 l of water per day were evaporated to increase the rH in the galleries. The room climate was measured to optimise control of the indoor conditions and in response to past bad experience. The documentation included an accurate register of the annual combustible material consumption, daily notes on water usage for evaporation, and records of daily rH and temperature measurements performed with a psychrometer.

5.1 Indoor Climate Conditions and Low-Pressure Steam-Heating

The results of the hygro-thermal simulations indicate the success. In the corse of one year (Fig. 6), temperatures never undercut the freezing point and only on some days fell below 10 °C. In most situations the relative humidity could be kept above 40%. During winter, relative humidity ranged between 40 and 60% with daily

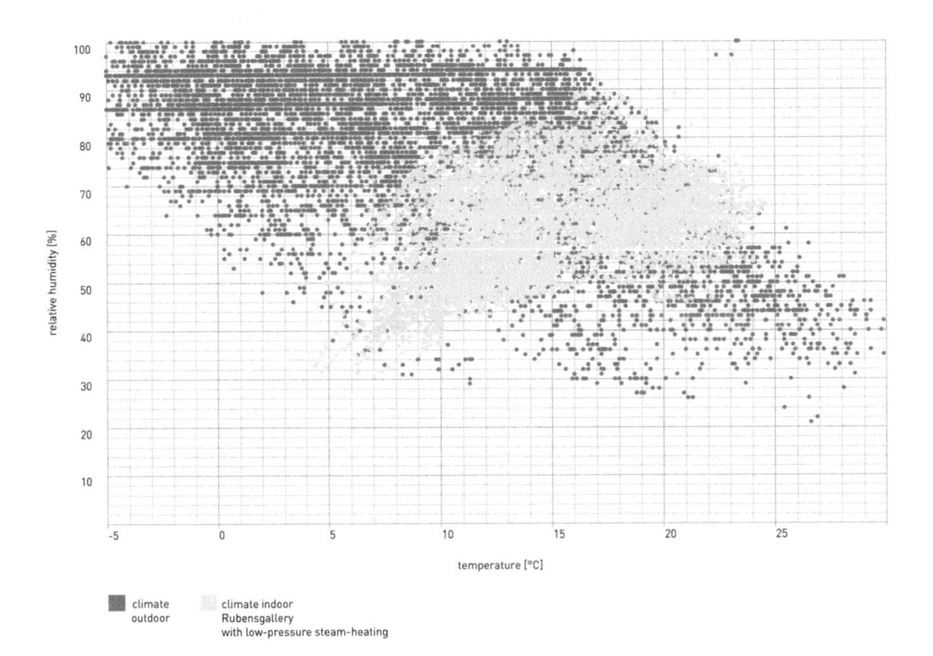

Fig. 6 Scatterplot of relative humidity and temperature over a period of one year during the time when the low-pressure steam-heating was installed (1891–1945)

fluctuations of about 5%. These could have been less, if a continuous heating would have been implemented instead of the postulated night setback. But finally, the low-pressure steam-heating was the best solution in terms of conservational aspects and technical feasibility.

6 From World War II to the Late 1990s

From 1939 onwards, large parts of the collections were evacuated, and the building was severely damaged by allied bombs in 1942 and 1944. Reconstruction of the building by the architect Hans Döllgast led to a massive alteration to Klenze's original architectural and technical concept.

The controversially discussed reconstruction campaign started in 1952 and lasted until 1957, when the museum re-opened. Reconstruction was conducted in three stages. It began with removal of debris, safeguarding of the roof and the bare brickwork. Döllgast treated the building's 'wounds' as a kind of memorial by leaving them openly visible. The interior fitting started in 1955, which again led to major changes in the original concept: the former main entrance on the east side was relocated to the north side in the centre of the building. The loggia corridor along the main galleries, which had been almost completely destroyed, was not reconstructed. Instead, it was turned into a hughe staircase connecting the entrance hall with the first floor galleries on both sides. The light lanterns were not rebuilt, but larger areas of the roof were glazed. In line with the reconstruction work, a simple air-conditioning system was installed which allowed the air to be heated and humidified. Cooling and dehumidification were not possible at that time. This led to high temperature and high rH in the galleries during the summer, which caused frequent complaints from visitors and security staff. The whole building was supplied with heat from various systems; the new entrance hall (former vestibule), for example, had floor heating. Only the galleries were completely air-conditioned at a rH of 60% and a temperature of 20 °C as minimum levels. However, the aerosol devices used for humidification of the warm air caused a substantial electrostatic charge of dust accumulating on the painting surfaces.

For the first time in the history of the Alte Pinakothek, artificial lighting was installed. Indirect illumination of the galleries was achieved by more than 900 fluorescent tubes with 64 and 40 W attached in double or triple rows to the cornice below the haunches. Furthermore, 16 electric bulbs with 300 W were also installed for cleaning purposes.[7]

[7]Staatsarchiv München LBÄ 2114, Cost estimate for the new arrangement of the Alte Pinakothek from 21 June 1956.

7 Overall Renovation in the 1990s

The disadvantages of the air-conditioning system dating to postwar period of the 1950s required a general refurbishment of the whole building. The museum was closed for four years between 1994 and 1998. Although the architectural concept of Döllgast's reconstruction was not touched in principle, a full HVAC system was installed. From now on, the air could be heated, cooled, humidified and dehumidified using 49 decentralised units supplying the gallery rooms, the depositories and the conservation studio. The air exchange rate is three to four times per hour. depending on the outdoor conditions, the fresh air rate in the galleries is about 10%. The lighting system now consists of 1543 fluorescent tubes with 36, 58 and 72 W, mirror reflectors and prismatic diffusors. The illumination concept combines artificial light and daylight.

A curtain system was installed in the attic to regulate incident daylight. The single glazing in the skylights was replaced by double glazing with UV filters. However, due to serious planning failures, the mechanical shading system had to be turned off some months after re-opening. To avoid high illumination levels, large textile sheets were permanently fitted to the skylights. Finally at the end of the 20th century, the Alte Pinakothek has turned from a daylight to an artificial lit museum.

8 Retrofitting and Energy Savings

The insufficient lighting concept and, above all, the low energy efficiency of the attic, recently prompted an exemplary refurbishment of one of the galleries (Gallery X). An overall concept for the energy-efficient retrofitting of the whole building has been developed on the basis of this gallery. The retrofitting started in 2014 and is still ongoing work. In Gallery X, the first stage was internal insulation of the roof using a layer of mineral wool, a vapour barrier and plasterboard. For the new glazing of the roof lights, a combined sun-protecting and heat-reflecting glass with a colour rendering index of 95% was chosen. The skylights were fitted with a light diffusing prism glass with a transmittance of 70%. To control incident daylight, a louvre system was installed in the attic. It is regulated according to the seasonal solar zenith angle. The artificial light installations in Gallery X were transferred into the attic. They consist of fluorescent tubes and four floodlights. However, this solution is not used for the retrofitting due to its high energy consumption. In practice the currently implemented lighting system is comparable to the post-war system.

9 The Energetic Price of Museum Climate

The hygro-thermal simulation of the different building phases aims to provide an initial picture of the historic conditions, namely rH and temperature as well as the particular energy flows necessary to maintain these conditions. Although the absolute values were not the main focus, the simulations allowed the indoor climate to be qualified and compared.

Scatter plots were used to facilitate comparisons of the simulation results for rH and temperature. These plots visualise the data over a period of one year (Fig. 7). A detailed description of the complex hygro-thermal simulation process as well as the results of light simulations can be found elsewhere (Bauernfeind 2016).

Not unexpectedly, every building phase shows a characteristic distribution of rH and temperature. In contrast to the post-war period, there were no technical measures to control the indoor climate between April and October in earlier times, when the building's envelope was the determining factor. Here, the massive masonry was responsible for the delayed and moderate influence of the outdoor climate on the indoor conditions. Between 1957 and 1994, the indoor climate conditions were kept above a temperature of 20 °C and 50% rH throughout the year. The main differences between phases 1–3 can be observed in the winter months, especially during the heating period. There was no temperature control at all during phase 2, and the indoor climate was characterised by low temperatures and high humidity. The

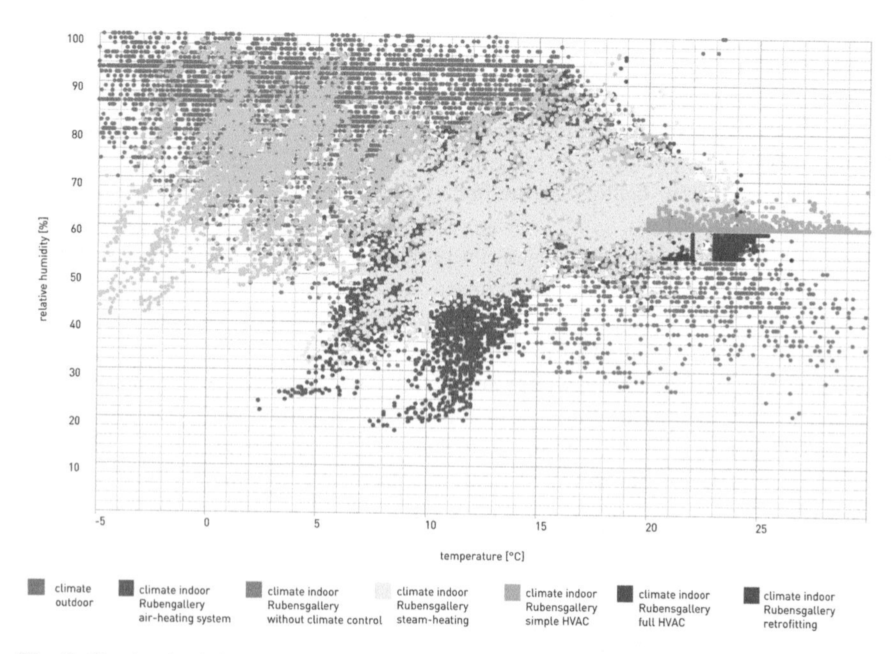

Fig. 7 Simulated relative humidity and temperature over a period of one year to compare phases 1–6

air-heating system was used to raise the temperature levels. However, serious humidity drops occurred. As we know, such drops provoked severe damage to the paintings on display. In consequence, the system was switched off. The low-pressure steam heating system was the first attempt to raise the temperature level and to humidify the air simultaneously. The sources reveal the important finding that during phases 1–3, temperature was not controlled for the sake of visitors but to keep the building above freezing. Preservation of the artworks was the main concern, or, as formulated by the commission in charge: 'the commission pointed out the significance of sufficient heating in its report, by emphasising that the temperature in every gallery has to be regulated in a way that condensation or wet deposits never occur on the paintings'.[8]

Due to Klenze's knowledge about conservation needs and the technical limitations of his time, his main intention was to keep the building above freezing. But the tenor during the planning of the low-pressure steam heating system at the end of the 19th century was quite different. The experts understood that rH is the most important factor for the preservation of artworks, and that temperature control alone was not sufficient. Moreover, they denied a further raise in temperature simply for human comfort, because this was associated with higher risks for the whole collection as well as drastic increases in energy consumption.

A detailed analysis of the climate fluctuations in phases 1–6 is shown in Table 2. It is obvious that the range of annual rH fluctuations became smaller from phase to phase. During phase 1, there is a deviation from the annual mean rH of about 40%. Heating during phase 1 narrowed the band of annual temperature fluctuations, whereas the highest annual fluctuation in temperature occurred during phase 2, when the building was not heated at all. During phases 1–3, the summer conditions seemed to have been fairly stable, whereas the highest daily rH fluctuations occurred during winter in phases 1 and 3. The reason why phase 2 shows lower fluctuations in temperature is the fact that the galleries were heated during the day

Table 2 Comparison of the annual and daily fluctuations of rH and T for the phases 1–6

Phase	Deviation from annual mean		Daily fluctuations winter		Daily fluctuations summer	
	Δ rH [%]	Δ T [K]	Δ rH [%]	Δ T [K]	Δ rH [%]	Δ T [K]
1	<40	<9	<10	<5	<5	<1
2	<32	<15	<7	<1	<5	<1
3	<26	<11	<10	<2	<5	<1
4	<12	<5	<2	<1	<5	<4
5	<5	<2	<2	<1	<2	<1
6	<4	<0.5	<1	<0.5	<1	<0.5

[8]See footnote 2.

during phases 1 and 3. Introduction of an air-conditioning system during phase 4 reduced the annual fluctuations. The daily fluctuations during winter were no longer of great relevance. In summer, the enlarged areas of glazed roof and the lack of cooling led to raised daily fluctuations in temperature. With the introduction of the full HVAC system in phase 5, the temperature and rH fluctuations could be minimised year round. A comparison of phases 5 and 6 shows that sufficient insulation of the roof and improved glazing achieved a further stabilisation of the indoor climate. This substantial improvement was reached without any changes to the air-conditioning system or the set point values for rH and temperature.

10 Past Knowledge for Future Perspectives

A simulation of the different phases in the building history of the Alte Pinakothek reveals that modifications to the building envelope or interventions in the climate control strategy have always influenced the indoor climate conditions. A comparison of the six investigated phases leads to an obvious conclusion: since the 1950s, the set points for rH and temperature have closely followed improvements of technical equipment for environmental control. The resulting uncontrolled increase in energy demand is now a double burden because both the financial considerations and the increased carbon footprint necessitate sustainable interventions. Lower energy consumption and an essentially stable indoor climate can be achieved through other means, as for example demonstrated in phase 3. To develop further successful approaches for the future, any museum building must be understood as a holistic system within its particular location.

Like the short review on climate control strategies in other museums shows, an amazing and nowadays quite underestimated knowledge on the correlation between indoor climate, visitors influence, climate control strategy and their influence on the state of preservation of artworks already existed in the 19th century. This knowledge as well as the exchange of experience between architects, museum directors and engineers also played an important role in the context of designing a new heating strategy for the Alte Pinakothek. Practice showed that the main focus had to be put on achieving a constant rH. The strived for conditions were not determined by human comfort but by preservation needs of the paintings. At that time, the first consistently climate measurements and recordings started in the Alte Pinakothek. Initially, rH and temperature were daily measured with a psychrometer. Later, continuous measurements with thermohygrographs and afterwards with digital sensors were performed. The fact, that these records still exist is an exception. In particular with regard to the hygro-thermal simulations this was a stroke of luck helping to verify the results.

Even though there is motivation and public pressure to juggle energy efficiency, cultural heritage preservation, preventive conservation and architecture, the way how this is objective is persued is questionable. Reducing the energy use by insolation and raising efficiency is one solution. But at the same time, the strategy of climate control with full HVAC systems and narrow set points for the whole building is to be revised. Is mechanisation the best principle for museum architecture? Is it possible to develop sustainable solutions combining adequate construction methods, prudent architectural concepts and systematically applied engineering? It should be kept in mind that sustainability does not solely mean energy saving. Sustainability is also the reliability and future perspective of any preservation strategy. This is why any sustainable solution must consider the energetic price to be paid for constant indoor climate conditions in museums.

Finally, the Alte Pinakothek and its rich collections turned out to be an excellent example of an holistic approach in museum architecture. Over most of its 175 years, the building has provided improved conditions for the preservation of its collections. This among others is a reason why this museum building has such a great impact on following generations of museums. Its patron, Ludwig I, from the beginning forced an interdisciplinary exchange between the architect Klenze and the museum director Dillis. Today, we would say that the museum's concept has been developed by a multidisciplinary team of architects, engineers and conservation professionals. Finally, learning from history is also an aspect of sustainability: Drawing conclusions from the past for tomorrow means implementing lessons learnt by earlier generations while considering the environmental, social and economic challenges for future generations that will arise from present-day decisions.

References

Anonym (1836) Beschreibung eines Luftheizofens, ausgeführt in dem neuen Königsbaue in München. Allgemeine Bauzeitung (8):57–60

Bauernfeind M (2016) Die Alte Pinakothek. Ein Museumsbau im Wandel der Zeit. Munich

Davies J, Ryder GV (1841) Ueber Perkins' Methode Gebäude mit heißem Wasser zu heizen. Polytechnisches J 81(LIV):209–215

Eibl M (2011) Lernen aus der Geschichte. Historisches Klima in Museen und resultierende Klimatisierungsstrategien am Beispiel der alten Pinakothek. Master thesis, Technical University, Munich (unpublished)

Frodl-Kraft E (1997) Gefährdetes Erbe. Österreichs Denkmalschutz und Denkmalpflege 1918–1945 im Prisma der Zeitgeschichte. Böhlau Verlag Wien, Vienna

Haag J (1873) Ueber die Beheizung der Personenwaggons bei den Einsenbahnen. Dingler's Polytechnisches J 207(CXVII):433–443 and plate IX

Heyl S (2012) Johannes Haag—Begründer der deutschen Zentralheizungsindustrie. In: Haus der Bayerischen Geschichte (ed) Industriekultur in Bayern, Edition Bayern, Sonderheft 5, Augsburg, pp 96–98

Maaz B (ed) (2001) Die Alte Nationalgalerie: Geschichte, Bau und Umbau. Berlin

Meier G (2011) Zur Archäologie der Haustechnik. In: Boschetti-Maradi A, Dieterich B, Frascoli L, Frey J, Meyer Y, Roth S (eds) Fund-Stücke—Spuren-Suche. Berlin, pp 571–591

Meissner PT (1827) Die Heizung mit erwärmter Luft erfunden, systematisch bearbeitet und als das wohlfeilste, bequemste, der Gesundheit zuträglichste, und zugleich die Feuersgefahr am meisten entfernende Mittel zur Erwärmung der Gebäude aller Art dargestellt und practisch nachgewiesen, Dritte, sehr vermehrte und gänzlich umgearbeitete Auflage. Wien

Peclèt JCE (1860) Vollständiges Handbuch über die Wärme und ihre Anwendung in den Künsten und Gewerben für Physiker, Berg-, Hütten, Fabriken- und Bau-Ingenieure, Mechaniker, Fabrikanten, Landwirthe etc., Bd. 1, Leipzig

Peclèt JCE (1986) Grundsätze der Feuerungskunde, Reprint der 3., gänzl. umgearb., sehr verm. u. verb. Aufl., Weimar 1858, Düsseldorf

Perkins AW (1841) Perkins über seine Heiflwasserheizung. Polytechnisches J 81(55):215–223

Schawe M (1994) Vor 50 Jahren—die Bayerischen Staatsgemäldesammlungen im Zweiten Weltkrieg. In: Bayerische Staatsgemäldesammlungen, annual report 1994, Munich

Schawe M (2010) 1947—die Bayerischen Staatsgemäldesammlungen. In: Lauterbach I (ed) Kunstgeschichte in München 1947. Institutionen und Personen im Wiederaufbau. Zentralinstitut für Kunstgeschichte, Munich

Voit E (1867) Die Neubauten im Königl. botanischen Garten in München. Zeitschrift für Bauwesen (17):316–324

von Dehn-Rotfelser H (1879) Das neue Gemäldegalerie-Gebäude zu Cassel. Zeitschrift für Bauwesen 29:9–34

von Klenze L (1930) Sammlung architektonischer Entwürfe, welche ausgeführt oder für die Ausführung entworfen wurden, Heft 1. Munich

Two Early Examples of Central Heating Systems in France During the 19th Century

Emmanuelle Gallo

Abstract Two early Examples of Central Heating Systems in France during the nineteenth century Emmanuelle Gallo Abstract In that paper, two case studies of heating history are detailed: the French stock exchange building in Paris the Palais Brongniart (1808–1826) and the Lariboisière hospital (1846–1854). With the first case, it was an opportunity to realize a diachronic approach, the evolution of several heating system and energy consumption in that very special program, from the first steam system in the country to the Parisian's district heating network since 1947. With the second case, we explained how a competition between two systems moved a building in use into a research laboratory with two different heat transfer fluid and two ventilation devices settled in the two opposite wings of the hospital. With those two cases, archive studies allowed to discovered new data on all sorts of energy used for heating and lighting since 1805 and their prices during the centuries. A new research program was the result of these investigations.

At the beginning, was my PhD, when I worked on the history of heating systems in France. At that time, I investigate more on inventions and inventors than on buildings. During that research, I discovered an incredible richness of archives about some public buildings erected and maintained during the last two centuries. I keep that in memory and when I had the opportunity, I get back to the archive of some buildings. It takes time because often documents are scattered in several different archives and it slowed the investigations. In this paper, I intend to present some examples of those researches, more or less detailed and how it opened new opportunities.

In previous papers, I have pointed out that inventions concerning heating were settled first for the production of goods and secondarily for the warming of public

E. Gallo (✉)
Brittany Architectural School, Rennes, France
e-mail: emmanuellegallo@free.fr

E. Gallo
l'AHTTEP (UMR AUSser CNRS 3329), Paris-La-Villette
Architectural School, 153 rue de Belleville, Paris 75019, France

C. Manfredi (ed.), *Addressing the Climate in Modern Age's Construction History*,
https://doi.org/10.1007/978-3-030-04465-7_4

buildings and even later for the domestic comfort (Gallo 2008; Gourlier 1825–1850). The nineteenth century is a period of creation and development of public buildings in France with city halls, post offices, schools, prisons, hospitals, theatre, train stations, court … At that time, it was possible, technically and financially, to plan or to introduce heating systems in those public buildings (new or not).

Working on those topics, I become aware that during the nineteenth century, we can see two different points of views: one for architects and another for engineers. In his book Eugène Peclet (1793–1857) *Nouveaux documents relatifs au chauffage et à la ventilation des établissements publics* listed public buildings case studies of interest to engineers (Peclet 1853). In an illustrated architectural book of Charles Gourlier *Choix d'édifices publics projetés et construits en France depuis le commencement du XIXe siècle* we can see another list of buildings of interest to architects. At the intersection of the two lists we found the stock exchange building, the Palais Brongniart, with his original rectangular-shape plan. It will be my first example, my second example will come from the engineer's list: the Lariboisère's hospital.

1 The Steam Heating System of the Palais Brongniart

The stock exchange's building decided by Napoléon in 1808, was initially designed by the architect Alexandre Brongniart (1739–1813) and after his death, Eloi Labarre (1764–1833) took over the job. They were helped by other younger architects and one was Charles Gourlier (1786–1857) in charge of the technical issues with the title of "inspecteur". The building is designed on a rectangular-shape plan (69 m by 41 m) laid on a base of 2.6 m, with a Corinthian column's gallery on the facades. The construction is characterized by the large central room measuring 37.60 m long and 24.68 m wide, with a height of 25 m, largely lit by an iron frame glass roof.[1] Other spaces in various sizes and functions occupied the space between the central hall and facades on two main levels. The building and its outdoor monumental staircase defines a large place in the West and contained, in addition to trading activities, the Paris's trade tribunal but also the labour exchange and commodities. In 1825, the initial intention was to install warm air stoves of different sizes, according to the importance of the space to be heated. After a committee was settled to avoid many chimney pots on the roof judged inappropriate for the majesty of the building. Louis Joseph Gay-Lussac (1778–1850, physicist), Louis Jacques Thénard (1777–1857, chemist) and Jean Pierre d'Arcet (1777–1844, chemist) member of that committee, have chosen steam heating because the large space at the centre of the stock exchange constitute an opportunity to try something new, something coming from England. So the steam heating system was designed for the main floor: the large room and the

[1]This iron frame glass roof, with a rectangular-shape is the second one in France of this size after the "Halle à Blé" iron frame. It is still in place and function.

offices around.[2] For this building, the steam boiler was made by a specialist, an English builder settled close to Paris: the *Forges de Charenton*, directed by Aaron Manby and Daniel Wilson (Belhoste 1988) (Figs. 1, 2, 3 and 4).

On the archives plans we can read that the steam boiler was installed at the South East corner, on the basement, in fact the natural ground level, underneath the main floor.[3] Several drawings of this heating system in the building were published in the *Bulletin de la Société d'Encouragement pour l'Industrie Nationale* and in Peclet's treatise, so we can figure the different devices and supply (*Bulletin* 1828[4]; Peclet 1828[5]). The network serving the stock exchange room was positioned underneath the corridor that encircles the central hall. The steam pipes, who could expand, were laid down in a gutter covered by cast perforated iron sheets and registers. The heat is transmitted by radiation or also rose through the air exchange. On the engraving, we could see some other heat exchangers were made of columns, building an inner façade around the corridor.[6] The other rooms on the main floor were heated with steam through heat exchangers looking like stoves. Residual waters were collected in boxes, before going back in the tank in the basement.

In published sources, it is written that there was just one boiler, but on different archives coloured drawings we can see two boilers.[7] Probably they used one boiler a few winters, and then added another one. May be it was more rational to warm different types of spaces, the large hall and the rooms around separately, because operating hours were different. The water tank, close to the boilers, collected rainwaters coming from the roof, but it was also connected to water available from Montmartre street.

With the elevation of the main floor, the original basement is located at the ground level, so it convenient to deliver coal and store it in the cellars by the East frontage. Wood and coal were also stored by the other institutions hosted in the building, who were warmed separately. On archives coloured drawings we can see different kinds of stock: wood or coal. We can also read drawings of chimneys on the main and second floors. Of course, the chimneys were not the only warming means in the other areas of the building that were not concerned with central

[2]We ignore what Brongniart have planned for the thermal comfort of the stock exchange's building, especially for the large room, because he died when the building site was at still the foundation's level. We may imagine that he could have planned something innovative. His son was at that time, a well known chemist directing the famous "Manufacture de Sévres", the Royal ceramics factory near Paris. That means that the Brongniart family had knowledge about heat production and regulation, ovens and boilers.

[3]Atlas 549.

[4]Tabb. 359–360.

[5]Tab. 117.

[6]We cannot be sure that this inner facade was existing as on the designs because it had been destroyed in 1903, when two wings were added to the building. Anyway, the idea to use decoration to distribute steam was very innovative.

[7]Atlas 549, archives de Paris and plans of the Bourse (before in a local archives, ares now unsorted in the Archives de Paris).

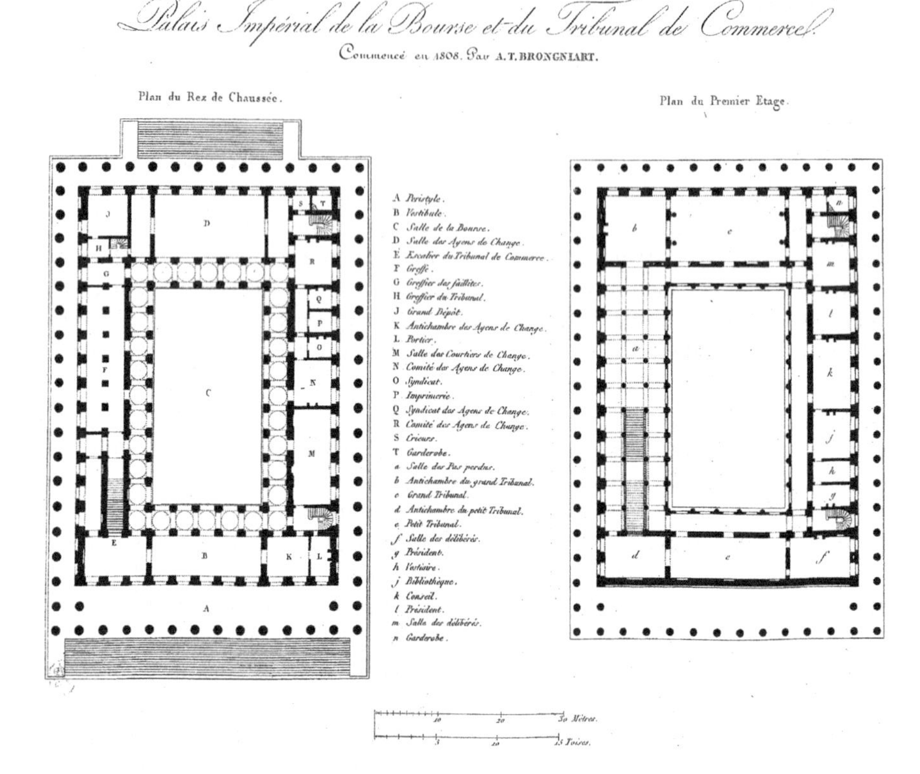

Fig. 1 The stock exchange building's plans in Alexandre Brongniart, *Plans du Palais de la Bourse et du cimetière du Mont-Louis, en six planches, précédées d'une notice sur les plans et quelques travaux du même artiste*, Crapelet, Paris, 1814, plate n°2

heating. They used large warm air stoves invented by Joseph François Désarnod and sold by the de Gernon's widow.[8] They bought different models, some rectangular, other cylindrical, covered with white tiles. The size, which could rose to three meters, depended on the importance of the space to be warmed (Figs. 5, 6, 7 and 8).

[8]Joseph-François Désarnod was an architect in Lyon when he tried to manufacture the Benjamin Franklin stove, in France. With this experience and some inventions he became one of the first warm air stoves and "calorifères" inventors in the country in the early nineteenth century. The de Gernon's widow succeeded Désarnod after his death and run the "manufacture royale en 1789 de Calorifères et Foyers salubres et économiques, 67 cour et passage des petites écuries faubourg poissonnière". At first, Charles Gourlier has ordered several big warm air stoves to Désarnod. But with the choice of steam heating, some large stoves were kept in stock. So, de Gernon's widow rioted because it was not easy to sell those big stoves to usual buyers. I found letters written by Charles Gourlier to the architects in charge of all the public buildings in Paris in order to find a place to adopt those abnormally huge stoves. Finally, they went to the *Ecole des beaux-arts* and in the *Halle à vins*, AN F/13/961.

Fig. 2 The stock exchange building in Alexandre Brongniart, *Plans du Palais de la Bourse et du cimetière du Mont-Louis, en six planches, précédées d'une notice sur les plans et quelques travaux du même artiste*, Crapelet, Paris, 1814, plate n°1

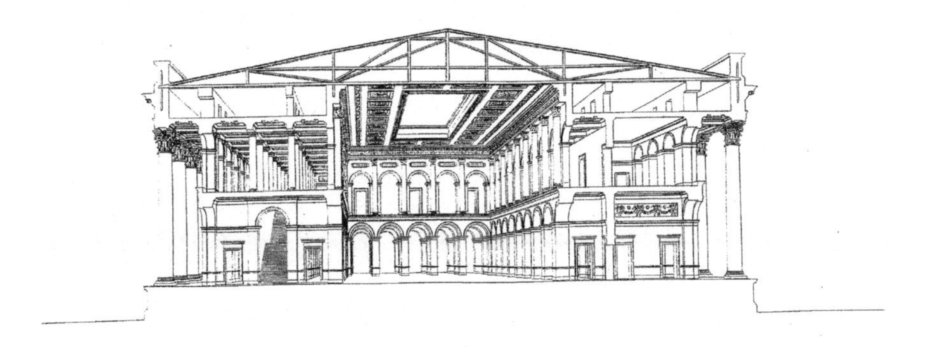

Fig. 3 Cut view of the stock exchange building in Alexandre Brongniart, *Plans du Palais de la Bourse et du cimetière du Mont-Louis, en six planches, précédées d'une notice sur les plans et quelques travaux du même artiste*, Crapelet, Paris, 1814, plate n°3

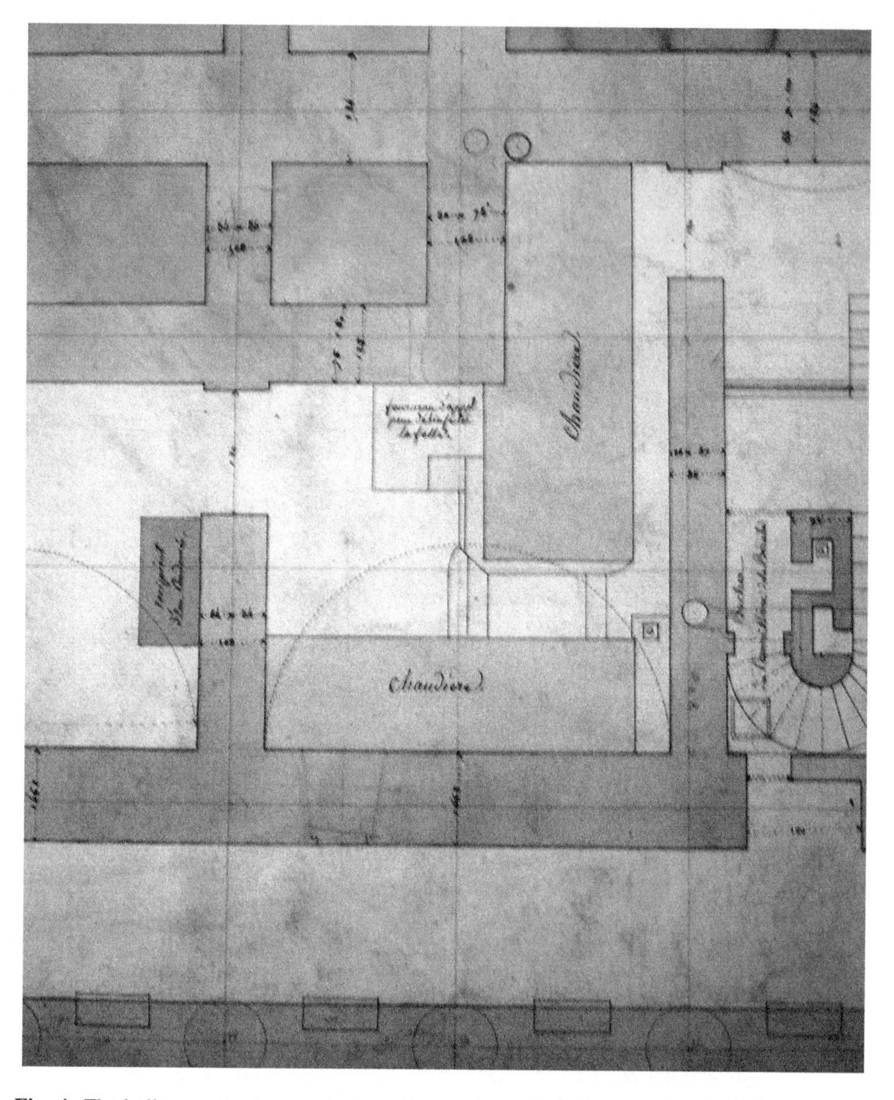

Fig. 4 The boilers on the basement plan, stock exchange building, archives de Paris

As the stock exchange was the first building heated by steam in France, the case was interesting to study. So, the stock exchange's offices recorded lot of information: how many hours the boiler in use every day, every week, every month, the costs of coal and oil … The records of the first two years were published in the *Bulletin de la Société d'Encouragement pour l'Industrie Nationale*, the principal newspaper for scientists of the time (Bulletin 1828[9]). During ten years, a lot of data

[9]P. 209.

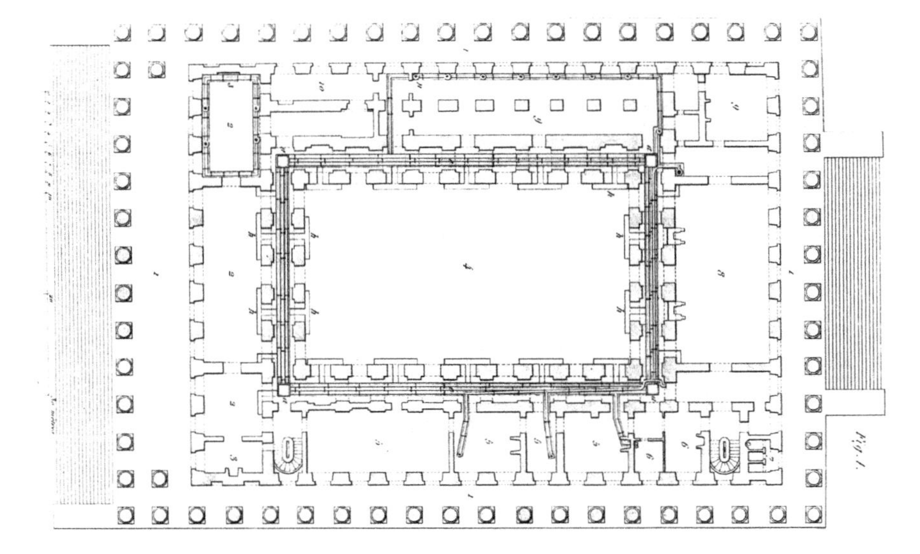

Fig. 5 The heating network, Peclet (1828, p. 117)

were recorded in handwritten booklets: how many hours with heating, how much coal was used, what kind of coal, how much it was costing and how many people were needed to use it.[10] The first years, the boilers drivers were employed by Manby and Wilson, they were well paid and very qualified, afterwards, the work was done by less qualified and less paid workers. The rather richness of the different archives allowed to discover so much, but quite a lot of drawings are difficult to date. Despite the difficulties, I can say that the steam heating system was successful enough to be extended to other areas where the warm air stoves were abandoned and relocated in other public buildings in town.

When I found in the archives all the records for ten years of activities, I asked myself if I could find this kind of data on other buildings. This discovery allowed another type of research, no more on inventions or how they were settled in buildings, but also how much energy was used and how much it costs to the society to warm a public building. It may allow us to compare the quantity and prices of coal and oil used for the heating and for the lighting of two buildings, for one or several periods … What was especially interesting about the stock exchange building study is that all the different discoveries have opened doors, and afterwards, it changes the research methodology on other buildings (Figs. 9, 10, 11 and 12).

[10]All the data may have been collected by the town, because the booklet is located in the Archives de Paris, AP VM27, box 9. The heating was usually started in October and stopped in April, as we do know.

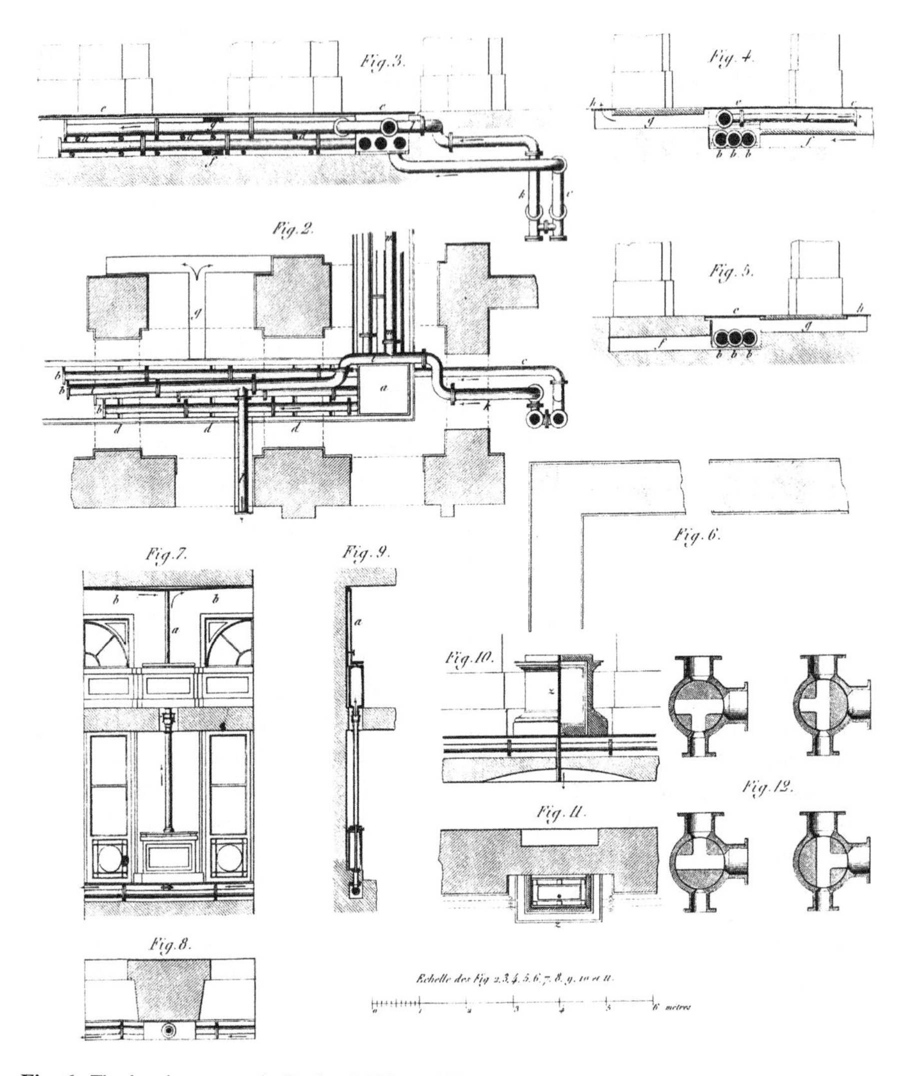

Fig. 6 The heating network, Peclet (1828, p. 117)

2 The Original Case of Lariboisière's Hospital

I have chosen another case to talk about, this time it is an issue concerning more engineers and people involved in the battle for hygiene: heating systems of Lariboisière's Hospital built in Paris in the middle of the nineteenth century. With this new hospital, the town and the *Conseil Général des Hôpitaux* intended to create an "exemplary" hospital for 600 patients. Located in the north of Paris, it's a symmetrical building around a courtyard with a church in the centre of the architectural composition. On both sides, one for women and another for men, the

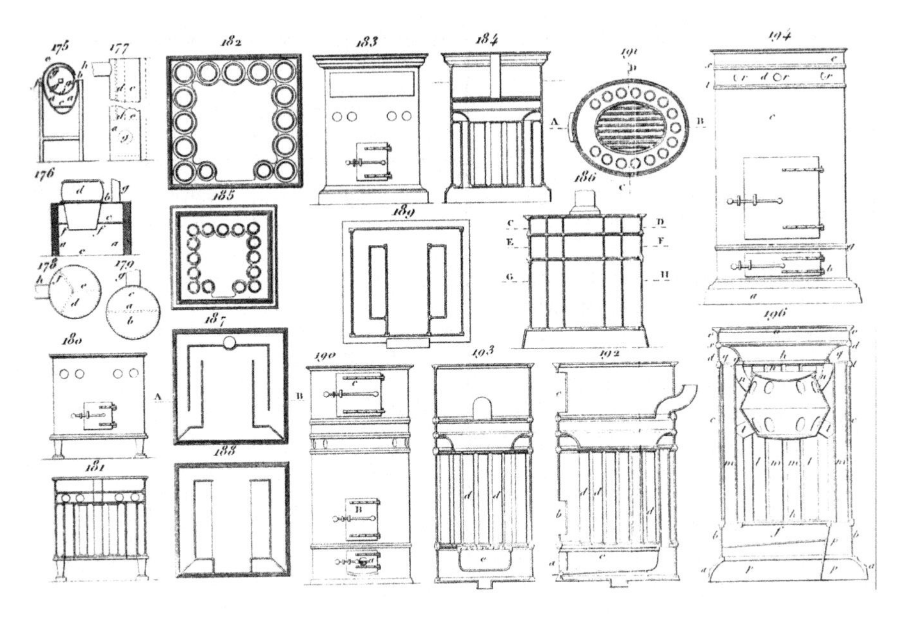

Fig. 7 Jean-François Désarnod warm air stoves, Ardenni, Julia de Fontelle, Malepeyre, *Manuel-Roret du Poêlier-Fumiste*, 1850, Roret, Paris, p. 8

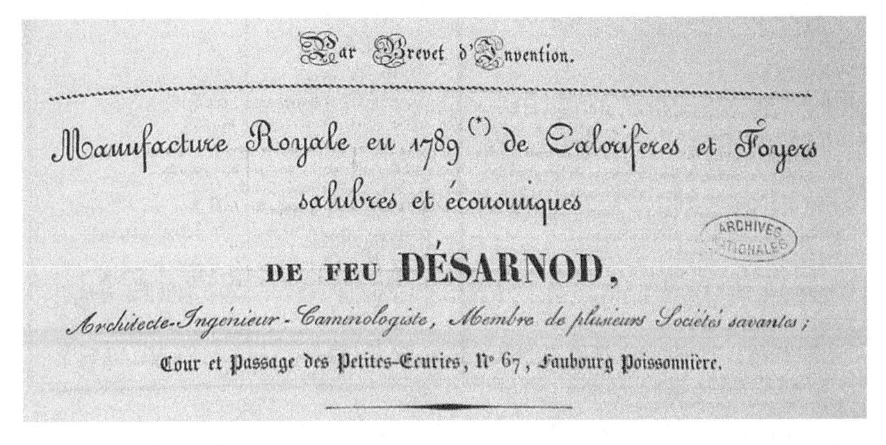

Par Brevet d'Invention.

Manufacture Royale en 1789 (*) de Calorifères et Foyers
salubres et économiques

DE FEU DÉSARNOD,

Architecte-Ingénieur-Caminologiste, Membre de plusieurs Sociétés savantes;

Cour et Passage des Petites-Écuries, N° 67, Faubourg Poissonnière.

Fig. 8 *Manufacture Royale en 1789 de Calorifères et Foyers salubres et économiques de feu Désarnod Architecte-Ingénieur-Caminologiste-Membre de plusieurs sociétés savantes*, 4 p., Archives Nationales

two-storey blocks were placed perpendicularly to covered corridors providing an efficient distribution. Each block contained three levels of thirty-two beds disposed in a large room and two additional beds in a smaller one. The architect, Martin Pierre Gauthier (1790–1855), trained by Charles Percier (1764–1838) and Pierre-François-Léonard Fontaine (1762–1853) was in charge of hospitals

	HIVERS	
	de 1826 à 1827.	de 1827 à 1828.
Date du commencement du chauffage. . .	1er novembre 1826.	3 novembre 1827.
— de la cessation du chauffage.	31 mai 1827.	30 avril 1828.
Nombre total des jours du chauffage, déduction faite des fêtes et dimanches. .	160.	149.
Heure à laquelle le feu a été journellement commencé.	de 5 à 7 h. du matin.	de 5 à 7 h. du matin.
Heure à laquelle on l'a cessé.	de midi à 3 heures.	de midi à 2 heures.
Nombre total d'heures pendant lequel le feu a eu lieu.	1350 1/2.	995 1/2.
Heures pendant lesquelles le chauffage a eu lieu,		
1°. Pour la grande salle.	de midi à 5 h. du soir.	de midi à 5 h. du soir.
2°. Pour les pièces accessoires.	de 9 h. du m. à 5 h. du s.	de 9 h. du m. à 5 h. du s.
Quantités consumées de charbon de Mons, première qualité et en pierre.	67,000 kilogr.	68,000 kilogr.
Prix.	75 fr. les 100 kilogr.	72 fr. les 100 kilogr.
Valeur totale.	5,025 fr.	4,896 fr.
Nombre de journées de chauffeur, compris nettoyage de l'appareil, réparations accidentelles, etc.	171 2/3.	125 3/4.
Prix.	4 fr.	4 fr.
Valeur.	686 fr. 66 c.	503 fr.
Dépense totale.	5,711 fr. 66 c.	5,399 fr.
Résultats moyens pour le chauffage de chaque jour:		
Durée du feu.	8 heures 44 cent.	6 heures 68 cent.
Charbon consumé, environ.	419 kilogr.	456 kilogr.
Valant.	31 fr. 43 c.	32 fr. 83 c.
Temps des chauffeurs.	1 jour 7 cent.	0 jour 84 cent.
Valant.	4 fr. 28 c.	3 fr. 36 c.
Valeur réunie du charbon et du temps. .	35 fr. 71 c.	36 fr. 19 c.

Fig. 9 Table, *Bulletin* (1828, p. 209)

construction and renovations from 1820 to 1855.[11] He followed Jacques-René Tenon's (1724–1816) ideals: "hospitals are tools or better machines to treat the sick".

[11]He had obtained the *Prix de Rome* in 1810 and taught at the *Ecole Polytechnique* succeeding to J. N. L. Durand.

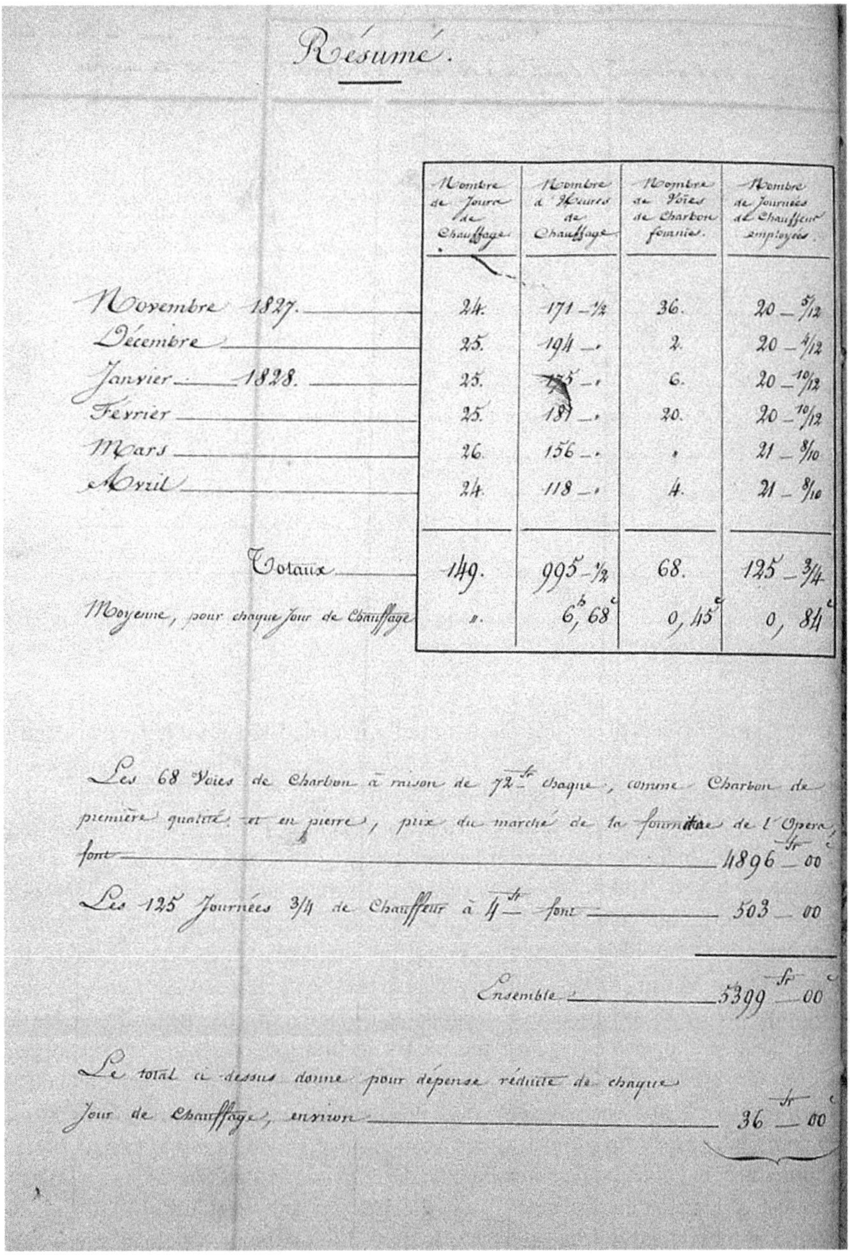

Résumé.

	Nombre de Jours de Chauffage	Nombre d'Heures de Chauffage	Nombre de Voies de Charbon fournies.	Nombre de Journées de Chauffeur employées.
Novembre 1827.	24.	171 – ½	36.	20 – 5/12
Décembre	25.	194 –	2.	20 – 4/12
Janvier 1828.	25.	175 –	6.	20 – 10/12
Février	25.	181 –	20.	20 – 10/12
Mars	26.	156 –	•	21 – 8/10
Avril	24.	118 –	4.	21 – 8/10
Totaux	149.	995 – ½	68.	125 – 3/4
Moyenne, pour chaque Jour de Chauffage	"	6h 68c	0, 45c	0, 84c

Les 68 Voies de Charbon à raison de 72fr chaque, comme Charbon de première qualité et en pierre, prix du marché de la fourniture de l'Opéra, font ———— 4896fr 00c

Les 125 Journées 3/4 de Chauffeur à 4fr font ———— 503 00

Ensemble ———— 5399fr 00c

Le total ci-dessus donne pour dépense réduite de chaque Jour de Chauffage, environ ———— 36fr 00c

Fig. 10 Energy data of the stock exchange 1827–1828, Archives de Paris, AP VM27, box 9

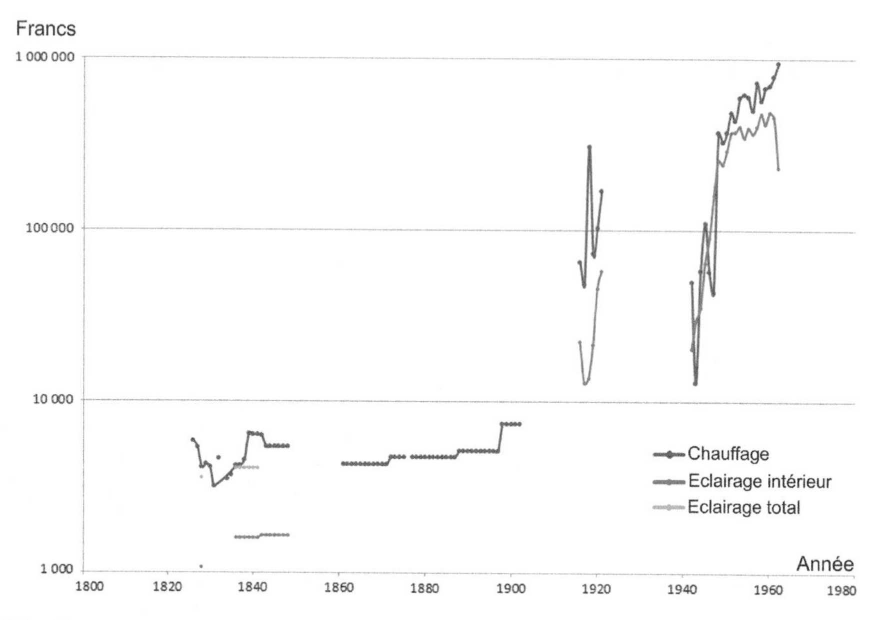

Fig. 11 Energy prices of the Stock exchange building, collected by *HPCE* team for the IMR *Ignis Mutat Res*

The *Conseil Général des Hôpitaux* initially intended to set up hot water central heating by Léon Duvoir-Leblanc.[12] This kind of system had been previously used, with good results, in older hospitals like Beaujon and Necker. But after the 1948 Revolution, *the Conseil Général des Hôpitaux* decided to organize a competition; with specified conditions: a uniform temperature of 15 °C and 20 m^2 of fresh air per hour for each bed. The engineers Grouvelle, Thomas and Laurens won the competition with the boiler builder Farcot. As the new design did not win the general approval and as Duvoir-Leblanc was protesting the decision, the council asked the general Arthur Morin (1795–1880), director of the *Ecole des Arts et Métiers*, to find a solution. With an experience in heating and ventilation, the general decided to divide the hospital into two parts: the men's section, one the right was equipped with the Grouvelle's steam system, the women's section, on the left, included the Duvoir-Leblanc's hot water system. With this singular decision the hospital became a research laboratory for almost fifteen years, where it was possible to experiment and measure in situ. It also generated a passionate debate between engineers, architects, doctors and pharmacists on the heating and ventilation of hospitals. Behind the theoretical debate, however, remained the issue of which system would control the lucrative market of the heating and ventilation of public buildings.

[12]The Léon Duvoir-Leblanc firm was well known for hot water heating systems named *calorifères* invented previously by Jean Simon Bonnemain (1743–1830), 24 rue Notre-Dame-des-Champs in Paris.

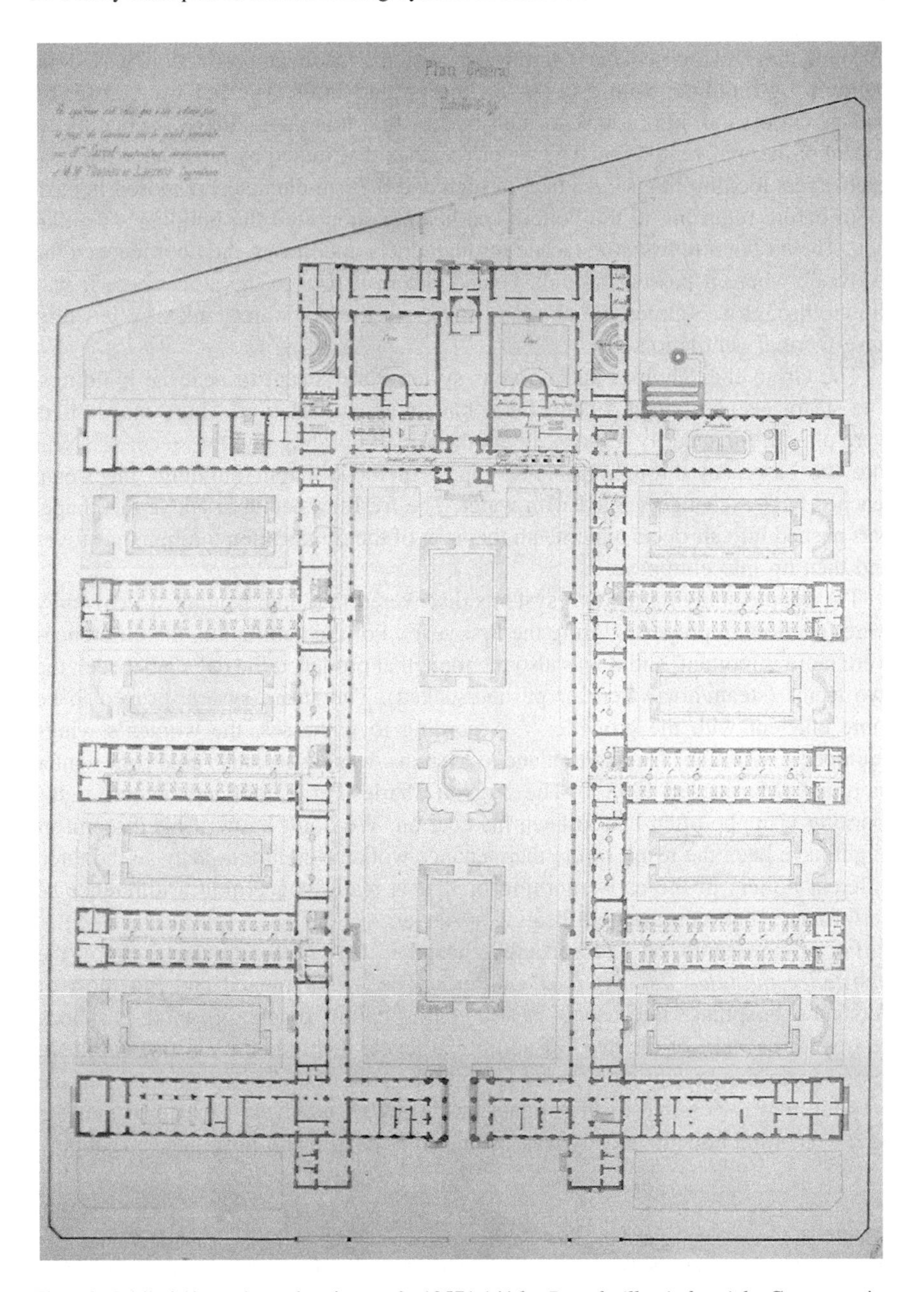

Fig. 12 Lariboisière plan, drawings n° 13571.1416. *Portefeuille industriel*, Conservatoire National des Arts et Métiers, Paris

With the Duvoir-Leblanc's hot water system, the heat production was done building by building, with a boiler on the ground floor. The hot water rose, by natural circulation, into a tank located in the attic; this huge tank was creating an appeal of thermal ventilation. The water was then distributed by pipes into four heat exchangers looking like stoves (1.5 m high and 0.79 m diameter) disposed in each room before returning to the boiler. The heating supported the building's ventilation. The air coming from the facade through ducts underneath the floor reached the "stoves", where it passed through, heated. Then, it rose to the attic where it was drawn through an impressive chimney stack, by the hot water tanks. Toilets also have thermal ventilation.

The Grouvelle, Thomas and Laurens system used steam to heat the buildings. The steam also activated a centrifugal fan for the ventilation. The steam boilers were placed, for security reasons, in another small building and the steam was sent into the basement through insulated pipes. In the different buildings the steam reached heat exchangers filled with water. The fresh air captured above the chapel was pushed into air ducts underneath the floor of the rooms, then through the stoves and then up into chimneys.

The people studying the two systems discovered that both heating systems they were functioning properly during the first years. For the ventilation, the differences were more important, and it was also the topic that produced the debate between the two teams (steam/hot water, air pushed/sucked). The steam system proved to be more efficient with the propeller.[13] According to witnesses, the women's wings equipped, with the Duvoir-Leblanc's system, was especially cold on winter mornings (Martineaud 1998[14]). The surgeon Charles Perier confided that during his morning visits he often had to keep his coat on. We don't know if his discomfort might have been the result of the maintenance workers' failure to activate the three different boilers in the early morning or if it is really concerning a difference of performance between the both heating systems.

For my research on Lariboisière's hospital, I went to the archives of the *Assistance publique des hopitaux de Paris*. The *APHP* owned and run most of Parisian's hospitals. Because it is still a hospital in operation, a lot has been destroyed, because of the need of additional levels (behinds the existing facades): more square meters and more beds! Despite the lack of archeological tracks in the hospital, it was possible to investigate in the *APHP* archives. I expected that the *APHP* recorded how the money was spent to run the different hospitals.[15] That how

---

[13]Both teams were fighting and the contributors wrote a lot of papers, mostly published in medical journals, and even a PhD on the topic. The debate lasted almost fifteen years; it slowed down with a deadly explosion that occurred in the *Saint-Sulpice* church in 1858, with a hot water system. Finally, the warm air heating won because hot water or steam could explode anyway; so they were not considered safe enough for public buildings.

[14]Pp. 42–43.

[15]Those archives have so many amazing data! For a while, during the nineteenth century in France, theatres had to pay taxes for the hospitals. So if you want to know how many people were going to a show you could know it through the *APHP* archives.

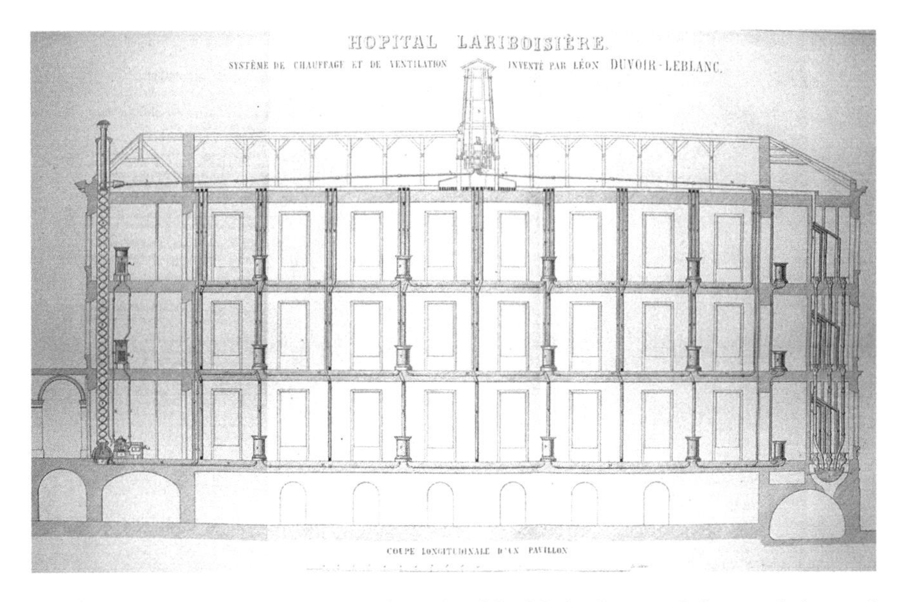

Fig. 13 Duvoir-Leblanc heating system in Arthur Jules Morin, *Rapport de la commission sur le chauffage et la ventilation du palais de Justice*, Paris, 1860

I discovered that I could found there: the price of wood, coal, oil, all sorts of energy, but also any usual goods, for every year since 1805. Thanks to the *APHP* archives, I was able to convert money into a quantity of energy and vice versa. Those data were essential to be able to compare the energy consumption of several buildings. In fact, for some buildings, I found tons of coal, and for others, you have some Francs per year. For public buildings, it is often possible to find accounting records, with lines dedicated to heating and lighting. So, thanks to those sources and after long hours of data gathering, it was possible to draw charts with the price of wood, coal, coke, electricity, gas, oil, ….[16](Figs. 13, 14, 15, 16 and 17).

[16]Those investigations took place in the project "Des profondeurs des caves à la canopée, histoire et prospectives des politiques énergétiques d'une capitale économe 1770–2050" or "From the depth of the Cellar to the Canopy, History and Prospective of the Energy Policy of Thrifty capital 1770–2050". I was directing the research team *HPCE* in charge of this issue, a winning contribution of the IMR *Ignis Mutat Res* call 2011.

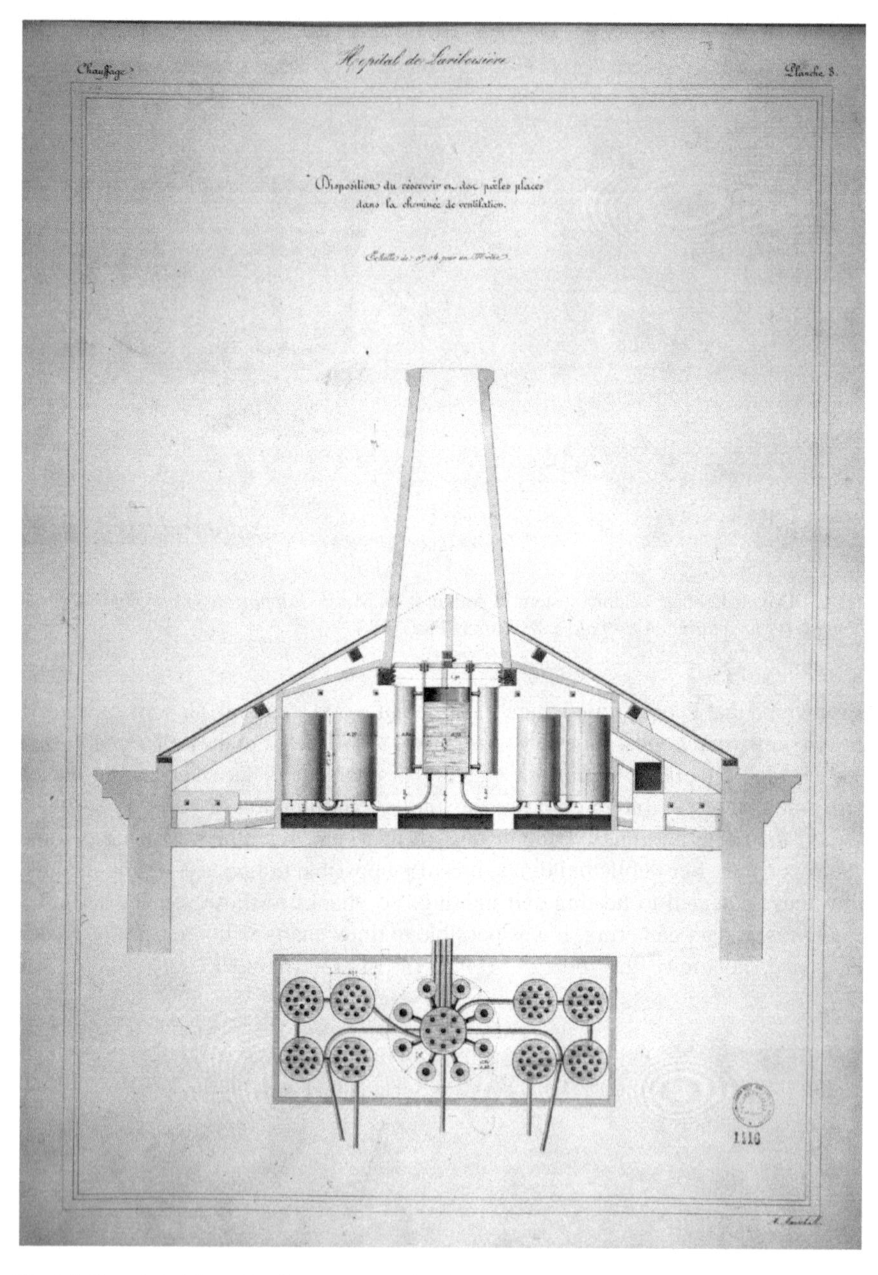

Fig. 14 Duvoir-Leblanc heating system, drawings n° 13571.1591. *Portefeuille industriel*, Conservatoire National des Arts et Métiers, Paris

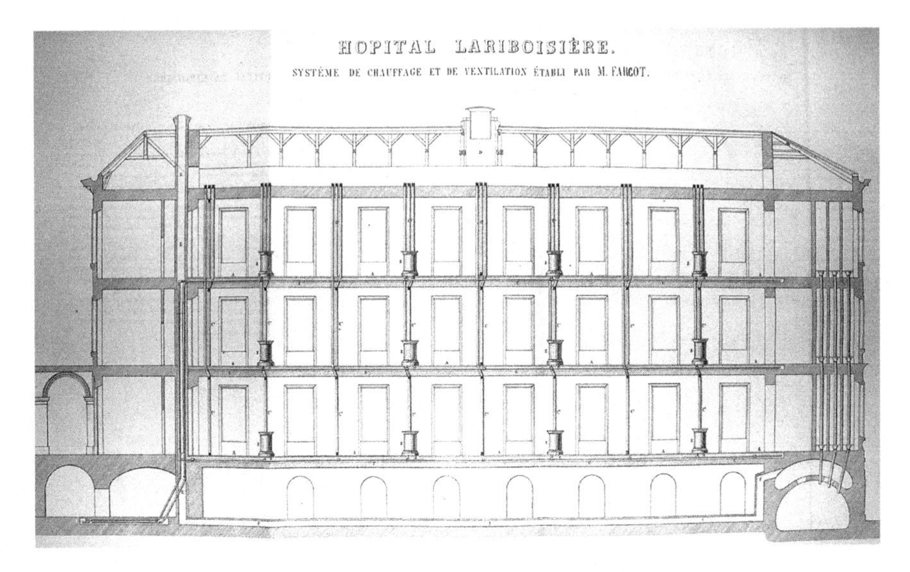

Fig. 15 Grouvelle (and Farcot) heating system in Arthur Jules Morin, *Rapport de la commission sur le chauffage et la ventilation du palais de Justice*, Paris, 1860

Art. 1. — Achats par les Établissements.

			Unité	Quantité	Prix	Total
Combustibles	Bois neuf	dans Paris	stères.	2,223 55	à 17.833	— 39.653 02
		hors Paris	—	1,200 07	à 15.693	— 18,832 72
		Exploitations	—	15	à 16.34	— 245 10
	Charbon de bois	dans Paris	hectol.	1,060	à 3.4041	— 3,608 45
		hors Paris	—	2.256	à 3,0464	— 6,872 80
	Houille. Gailleterie.	Forges	kil.	62.000	à 0.0318	— 1,973 40
		La Roche Guyon	—	»	à »	— » »
		Berck	—	499,670	à 0.0233	— 11,642 31
	Houille. Tout-venant	hors Paris	—	»	à »	— » »
	Houille. Charbon pour forge		—	22,500	à 0.0566	— 1,275 50
	Coke	dans Paris	hectol.	104,795	à 1.4668	— 153.719 24
		hors Paris	—	19,990	à 1.1683	— 23,354 80
	Margotins	commerce	Nomb.	319,200	à 0.0671	— 21.123 50
		exploitations	—	37,630	à 0.0642	— 2,417 93
	Braise		hectol.	472	à 2.0552	
Frais accessoires et indemnités	Frais de mesurage, de sciage (journées et menues dépenses)					
	Remboursement à des Administrés des Ménages de la valeur des combustibles non livrés en nature					
	Personnel professionnel fixe, 55	9 mécaniciens et aides				
		46 chauffeurs et aides				
	Gratifications					
	Indemnités	au médecin de Forges-les-Bains				150 40
		au Directeur de Cochin				150 40
		au Directeur de Broussais				200 40
		à 9 Pharmaciens à 150 fr.				1,350 »
		à divers Sous-Employés logés en ville				3,226 66
Combustibles	Gaz					
	Huile à brûler, hors Paris		kil.	25,109 80	à 0.6456	
	Chandelle		—	76 725	à 0.8800	
	Bougie stéarique		paquets	2,708 »	à 0.8161	

Fig. 16 Example of energy prices in *APHP* records *Comptes Financiers*, 1890, p. 96

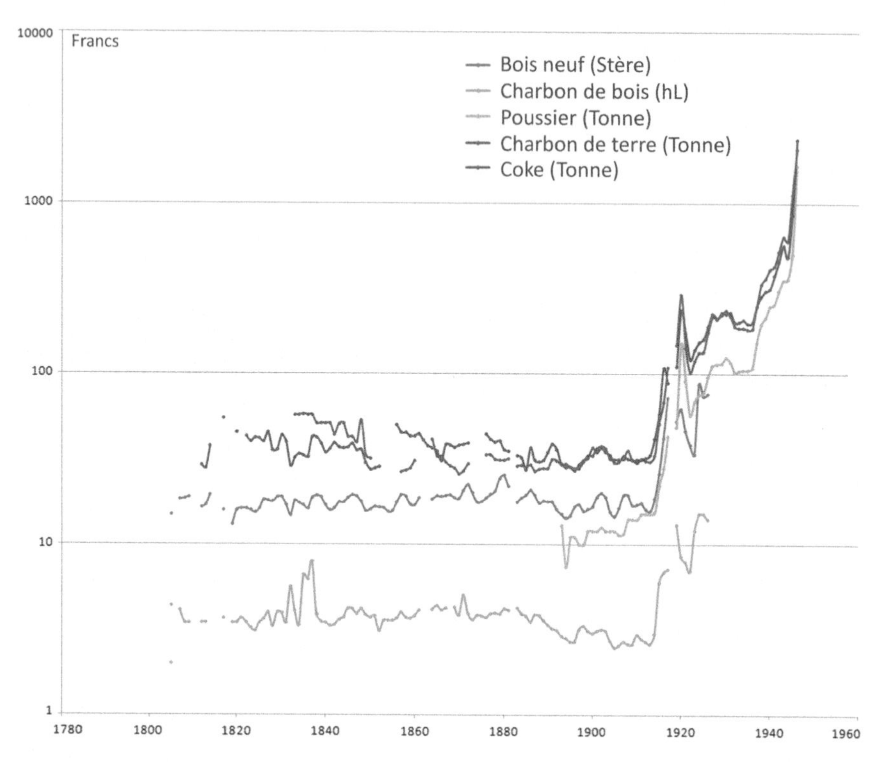

Fig. 17 Some energy prices in *APHP* records by *HPCE* team for the IMR *Ignis Mutat Res*

3 Conclusion

Of course, those early heating systems were renovated several times and later the heating system change more radically. In both cases, the buildings have also changed radically: the stock exchange was extended with two wings in 1903 and the hospital rebuilt a new structure inside the nineteenth century façade (during the second part of the twentieth century). It is sometimes possible, for public buildings, to follow those transformations because contracts and technical reports were kept and often finished in archives.

Through these examples, I showed how I made my research move from the investigation concerning early heating systems in buildings, to the cost of energy in Paris and energy consumption of public buildings in the capital. The archives sources richness allows the opening of new doors. Jeanne Singer-Kerel with her book published in 1961: *Le coût de la vie à Paris de 1840 à 1954* also used the *APHP* sources (Singer-Kerel 1961). The next steps may be study of the large quantity of different sources of energy quoted in those sources: new wood, drift-wood, washed wood, disused wood paving, brushwood, “margotins”, charcoal, coals, anthracite coal, coking coal, coke breeze, coke dust, shale gas, brown coal briquette or charcoal briquette, peat, coal peat, gas, vegetable fuel oil, Quinquet's fuel oil, lamppost fuel oil, natural candles, lanterns, industrial candles, gas lighting, oil, lamp oil, denatured alcohol, electricity….

References

Archives Nationales (AN) F/13/961

Archives de Paris (AP), VM27, box 9

Atlas 549, Archives de Paris

Belhoste J-F (1988) «Les Forges de Charenton», Architectures d'usine en Val-de-Marne (1822–1939). Cahier de l'inventaire 12:22–32

Gallo E (2008) "The lessons that can be drawn from the history of heating regarding the present situation", from a French perspective. In: The culture of energy. Cambridge Scholars Publishing, Newcastle, pp 268–280

Gourlier C-P (1825–1850) Choix d'édifices publics projetés et construits en France depuis le commencement du XIXe siècle, 3 vol. Louis Colas, Paris

Martineaud J-P (1998) Histoire de l'hôpital Lariboisière ou le "Versailles de la Misère". L'harmatan, Paris

Notice sur le chauffage par le moyen de la vapeur et application de ce procédé au chauffage de la grande salle et de plusieurs autres parties de la Bourse (1828) Bulletin de la Société d'Encouragement pour l'Industrie Nationale, vol 27, Paris

Peclet E (1828) Traités de la chaleur et de ses applications aux arts et manufactures. Malher, Paris

Peclet E (1853) Nouveaux documents relatifs au chauffage et à la ventilation des établissements publics. Hachette, Paris

Singer-Kerel J (1961) Le coût de la vie à Paris de 1840 à 1954. A. Colin, Paris

Camillo Boito and the School Buildings Indoor Climate in the Unified Italy (1870–1890)

Alberto Grimoldi and Angelo Giuseppe Landi

Abstract Compulsory schooling in some Italian States, especially in the Duchy of Milan, dated back to the end of the eighteenth century. Following the legislative proposal submitted by Minister Michele Coppino, compulsory schooling in the whole new kingdom of Italy initiated instead only around 1877, over 16 years later than the unification. Even if the Italian State funding were low, they nevertheless encouraged the construction of new school buildings and imposed new standards. Italian architects showed a great interest in the topic and develop new solutions, sometimes inspired by current European experimentations and publications. Architect Camillo Boito built two schools, both in Padua (1877–80) and Milan (1886–1890), which represented a concrete example of his idea of a new and "national" architecture. Up-to-date construction techniques were developed and thermal comfort was also improved, according to the standards of that period. The centralized heating systems adopted in the two school buildings testified to the fast development of industry, particularly in Milan, which become the most important Italian center in the sector.

1 Introduction. From Mandatory Schooling to School Buildings; from the Age of Enlightenment to the Belle Époque[1]

Compulsory education grew up slowly in nineteenth-century Europe, creating new issues of great significance, including of a quantitative nature, to the sector of public building, the schools themselves in its various types.

[1]This research develops and completes a lecture held at the 9th International Conference on Structural Analysis of Historical Constructions (Mexico City, 14–17 October 2014). This paper is the outcome of a common research by Alberto Grimoldi (about the biographical aspects) and Angelo Giuseppe Landi (institutional aspects, buildings and case studies).

A. Grimoldi (✉) · A. G. Landi
DAStU—Department of Architecture and Urban Studies, Politecnico di Milano, Milan, Italy
e-mail: alberto.grimoldi@polimi.it

C. Manfredi (ed.), *Addressing the Climate in Modern Age's Construction History*,
https://doi.org/10.1007/978-3-030-04465-7_5

In terms of quantity, rural schools were certainly at the heart of the problem; indeed, in the small towns and in the cities of Catholic countries, after the abolition of religious orders and brotherhoods, the ample spaces of monasteries and the halls of the confraternities could be assigned to the education. France had precociously decreed compulsory and free primary education in 1793, only to quickly abandon it. In the Habsburg States, as early as the final years of the reign of Joseph II (in Milan in 1788), the matter proceeded with great difficulty yet did not founder: a kind of construction manual, Koller's *Der Praktische Baubeamter*, presented not only models for rural schools but also a regulatory framework (Koller 1800).[2]

Certain characteristics relating to health were already emerging: lighting of desks had to be from left to right. With the advance of industrialisation, school buildings as a building type, in terms of their overall function and educational impact, as a design topic, took on particular importance. The regional imbalances increased with the industrial take-off, and combined with the population increase; re-use of existing buildings no longer sufficed and, with new buildings now considered indispensable, specific standards were progressively defined (Fig. 1).

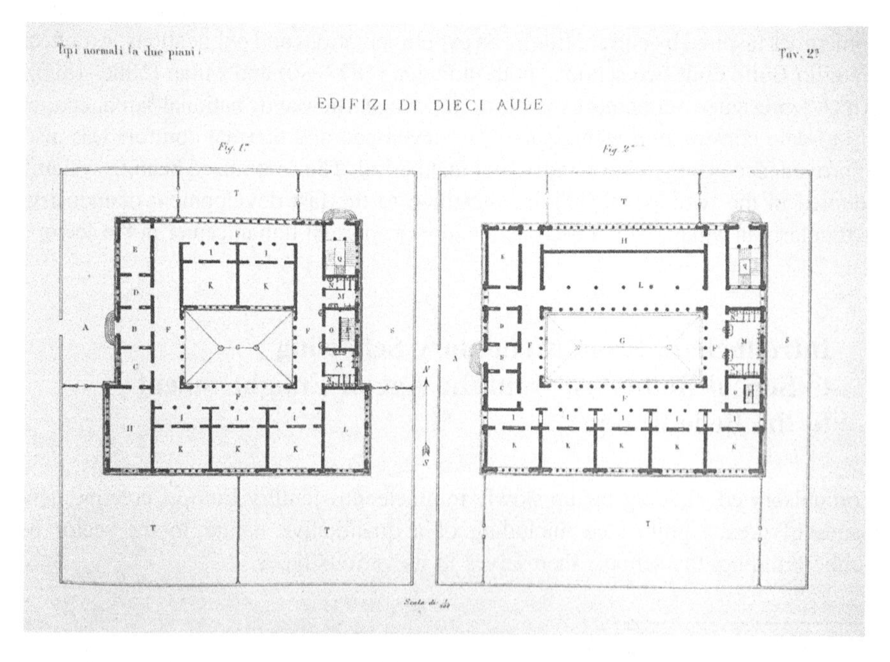

Fig. 1 School of ten classrooms ("Edifizi di dieci aule, planimetrie"), Bongioannini (1879), tav. 2

[2]pp. 39–41, figg. LXXVI–LXXVIII; II part, pp. 410, 458–461.

From 1851, the World Fairs, whose purposes included education, dedicated specific spaces to school buildings. Legislation reinforced public schooling and, after a century of attempts, mandatory primary education began to be effectively translated into practice and its duration was increased to three years. The education laws of the German States, that of the Habsburg Empire of 1869, the Education Act of 1870 in England and 1872 in Scotland and the laws of 1877 in Italy and 1882 in France boosted significant building activity and gave rise to regulations governing it. Specific bibliography on school building health grew notably, and crossed citations became numerous.

An overview of the history of Italian school building was still lacking, while literature on schools as an institution and on teaching was extensive although uneven. However, the previously-mentioned bibliography at European level, together with certain fortuitous circumstances, permitted, between the Italian Unification and the end of the nineteenth century, an overall vision of one particular aspect: climate control within schools. In this sense, the general concept of the buildings was still more determining. Subsequently, after the first world war, ventilation would become a secondary exigence in reference to the new medical guidelines, and the industrial take-off and diffusion of electric motors would reduce the bulk of installations, entrusting ambient balance to them.

2 School Legislation in Italy After the Unification and the Birth of Regulations on School Buildings

The Casati Law (1859) required municipalities to institute and finance the first two years of primary education in all areas where there were at least fifty children, but subordinated this to the requirements of the population, as determined by the municipal councils. In 1877, Education Minister Michele Coppino reduced compliance to one school per municipality, but made it possible for the Government to substitute for the defaulting municipalities at their expense. In return, he provided meagre financial support; indeed, Law no. 4460 of 18/07/1878 granted municipalities the faculty to contract loans of up to thirty years with the Italian Deposits and Loans Fund—in part with interest limited to 2%, thanks to the State's contribution—for building or repair of school buildings. This Law limited the interest burden on the Italian Ministry of Public Education to 50,000 lire per year. This assistance consisted, more than anything, of the offer of sufficiently ample long-term funding at a reduced cost (De Fort 1979),[3] but had limited results (Fig. 2).

The implementing regulation[4] defined the first brief guidelines on building choices. The substantial reiteration of these contributions at the end of the decade[5]

[3]pp. 185–186.

[4]Regio Decreto no. 4684, 13/12/1878 (Gazzetta Ufficiale 24/2/1879, no. 45).

[5]Law no. 5616, 8/7/1888 (Gazzetta Ufficiale 16/7/1888, no. 167).

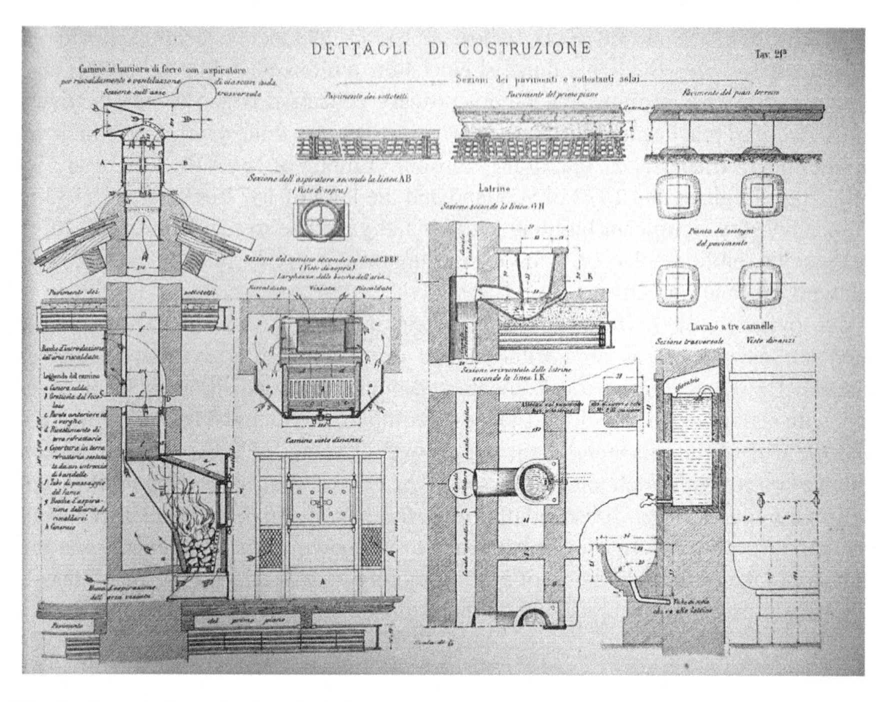

Fig. 2 Construction details, including the heating system ("Dettagli di costruzione"), Bongioannini (1879), tav. 21

was accompanied by more detailed instructions.[6] The new century began with a further reiteration,[7] together with new regulations. This sequence of events shaped increasingly detailed regulations on school buildings,[8] the three pages and 13 articles of 1888 became five pages and 18 articles in 1901. The relationship of these with buildings, constructed or merely designed and published, was two-fold. The "instructions" adopted the introduction of certain advanced solutions which had emerged during the previous decade, and their enactment was also notable in parallel changes to some standard dimensions or solutions. Overall, rapid technical progress was perceivable together with a change in expectations and attitudes.

For instance, in 1888, the required artificial lighting consisted of lamps situated at least 1.5 m from the students' heads. Where electric lighting is not used, oil, petroleum or gas burners should be chosen. Where gas lighting was adopted, the application of combustion product waste pipes was recommended. In 1901, an entire article recommended the use of incandescent electric lighting. If it was

[6]Regio Decreto no. 5808, 11/11/1888 (Gazzetta Ufficiale 30/11/1888, no. 282), "Regolamento ed istruzioni tecnico-igieniche per l'esecuzione della legge 8 luglio 1888 sugli edifici scolastici".

[7]Law no. 260, 15/7/1900, (Gazzetta Ufficiale 21/7/1900, no. 169).

[8]Regio Decreto no. 484, 25/11/1900 (Gazzetta Ufficiale 10/1/1901, no. 8).

necessary to resort to gaslight, each lamp should be fitted with incandescent mantles. Electric lighting quickly spread in the major cities, where the demand for new schools was highest, while energy transmission made it increasingly widespread. Gas lighting, dependent on supply networks, was used in a limited number of urban centres and, in 1901, only its most advanced and low-consumption form—the Auer von Welsbach gas mantle, patented in 1876—merited the regulation drafter's attention.

However, rural schools kept using oil (or petroleum) lamps with glass flues ("waste pipes"), despite the alternative of "combustible gas" (namely acetylene) proposed by the regulation of 1901.

3 The Museum of Teaching and Education

In reality, as would occur the following year with regard to financing, Coppino rationalised or developed earlier initiatives and brought them to fruition. These were the projects of former Education Ministers Cesare Correnti (1872) and Antonio Scialoja (1873). The need for improvement in the quality of education was generally felt and Ruggiero Bonghi,[9] Minister of Public Education from September 1874 to 1876, founded the Museum of Teaching and Education[10] in 1874 with this objective. Following a trend that was spreading throughout Europe (Tauro 1903[11]; Cossetto 1997), this Museum gathered information and educational materials, from books to wide-ranging objects, in the service of that teaching by means of things, "objective lessons", models or samples, that distinguished the nineteenth-century education. The Museum aimed to become a reference point for the schools of the Kingdom of Italy, and for this objective published its own journal (Sanzo 2017[12]). Bonghi announced the programme in his report on the Vienna World Fair in 1873 (Bonghi 1873[13]), and the same book featured a report on architecture by the architect Camillo Boito (Boito 1873[14]). School buildings, naturally, formed a—limited but essential—part of this programme.

[9]From 1866 to 1874 he was editor of the Milanese daily newspaper *La Perseveranza*, which published some essays of the architect Camillo Boito [from 1860 to 1865, according to the list of his publications in s.a., Boito (1916, pp. 176–205)].

[10]Regio Decreto 15/11/1874 no. 2212, published by Tauro (1903, pp. 9–10). The most relevant documents on the foundation and the live of the Museum life are collected by Sanzo (2012b). Concerning the object of this paper, a nearly contemporary and already mentioned witness, Giacomo Tauro and his work, is very useful. For a more overall perspective Sanzo (2012a) and also Meda (2010).

[11]pp. 3–5.

[12]pp. 99–120, antecedent bibliography pp. 116–119.

[13]pp. 33–34.

[14]pp. 17–29.

Through a Ministerial Decree of 30/12/1875, Bonghi established the commissions that would oversee the Museum's activities, including the one for school building design, composed of a doctor (Angelo Scarenzio), an agronomist (Fausto Sestini) and two engineer/architects (Francesco Bongioannini and Giulio de Angelis). The latter is very well-known, being among Rome's most original late nineteenth-century architects and, at that time, he had just designed a villa in Rome for the Minister Bonghi (Zullo 2005[15]); after a successful professional career, he returned to public employment in 1895, as Director of Lazio's Monuments Department (Zullo 2005[16]; Miano 1987).

4 Francesco Bongioannini

The former had been an inspection engineer at the *Soprintendenza archeologica* of Rome from 1872 (La Rosa 2011[17]; Scavi 1872[18]) and, when this was closed, in 1875, he entered the newly-established *Direzione Centrale per gli Scavi e i Musei* (Melis Tosatti 2005[19]) in the capacity of topographical engineer; the following year (La Rosa 2011[20]) he became General Inspector for Architecture, preferred to Giulio de Angelis.

In this way, he moved into the "Divisione 2a" which, from 1877, assumed the name *Provveditorato alle Arti* of the Ministry of Public Education, directed by Giulio Rezasco (La Rosa 2011[21]). When the *Provveditorato* and the *Direzione Centrale per gli Scavi e i Musei* were merged by the Minister Guido Baccelli, in 1881, Bongioannini continued his work and, when in 1887 the direction was structured in three sections he assumed directorship of the "Divisione II", with responsibility for conservation of monuments (Melis Tosatti 1999[22]). After the restructuring ordered by Minister Pasquale Villari, in 1891, Guido Baccelli, returning as Minister in 1895, partially re-established the *status quo* but Bongioannini returned to Turin as superintendent of schools (L'osservatore scolastico 1896[23]), and, at the end of his career, was transferred to Alessandria, in the Piedmont region (I diritti della Scuola 1908[24]). He can again devote himself to the school buildings (Bongioannini 1898).

Bongioannini's role at the Ministry of Public Education made him the immediate administrative reference person for Camillo Boito, a member of the *Giunta*

[15]pp. 38–39.

[16]pp. 144–190.

[17]p. 25.

[18]p. 34, no. 47.

[19]pp. 186–87.

[20]p. 27.

[21]p. 31.

[22]p. 190.

[23]p. 12 (nomination: 26/8/1896).

[24]p. 531.

Superiore di Belle Arti since 1879. This relationship may be deduced from the murky and inglorious affair of the destruction of the Church of San Giovanni in Conca, in Milan, an obstacle to the straight stretch of the new via Carlo Alberto, in which the two cooperated to promote the City Council's demolition intentions (Colombo 2005[25]).

This singularly close relationship between the debate on school buildings and the care of monuments is of little wonder: both competences were centralised in a single ministry, often managed with equal interest by a single minister, pointing more to a network of personal relationships than to a continuity and sharing of concepts.

Bongioannini's texts place greater emphasis on the topics of education, although their titles are, at times, deceptive. The multiplicity—to the point of confusion—of his interests and their interweaving with his administrative activity are typical of that century, thanks to a population curve that quickly led to positions of responsibility. Fluctuation between utopia and mania and between egocentricity and social commitment are characteristic of the nineteenth century, meaning that both actions and texts should only be assessed with close reference to their context, taking no account of the different meanings that words and phrases may subsequently have assumed later. Another great and polyhedric bureaucrat of that era, Carlo Alberto Pisani Dossi (Melis 1996[26]; Serra 1987), depicted this environment and these tendencies—linking society and architecture—in a famous pamphlet dedicated to the "madmen" in the competition for design of the national monument to Vittorio Emanuele II in Rome, called "Vittoriano" (Pisani Dossi 1884). The tone he adopted —exasperatedly grotesque—highlights the tendencies and characteristics of an era, which permeated its daily life but in that dimension they appear attenuated as with inactivated vaccines.

Bongioannini dedicated his engineering degree thesis to heating (Bongioannini 1870), another key theme in the problems discussed here, and initially he planned to base his professional identity on these issues. The habit, that time, of the *Scuola di applicazione per gli Ingegneri* in Turin, of presenting degree theses as printed pamphlets made his brief essay easily accessible. In Rome, the Piedmontese origin —or the exile to Turin between 1848 and 1859—of a significant portion of the higher bureaucratic levels, provided the young engineer with a promising network of relationships, and his sufficiently generalist technical expertise permitted him to meet a varied demand. He was able to achieve the drainage of waters in the Roman Forum, in order to resume excavations (Scavi 1872), but also to design the layout for the arrangement of the *Collegio Romano* as Roman seat of the National Library, one of Ruggiero Bonghi's priorities. Like de Angelis with Bonghi, he designed the Roman residence of Minister Michele Coppino, in 1886: a five-storey rental house at via Cavour no. 194 (La Rosa 2011[27]). The abstract regularity of the fascias and stringcourses, which show the identical height of the internal spaces, allude to an

[25]Overall pp. 96–97.

[26]pp. 169–170.

[27]pp. 51–52 and 68.

only partially-existing structure. Any revival or adaptation of Renaissance themes to new proportions and requirements, so widespread in the Rome of his day, is distant, and this anomaly also attracted Paolo Portoghesi, now sixty years ago (Portoghesi s.d.[28]). While unable to establish its paternity, he connected it to a limited but significant group of buildings which could refer to similar cultural roots. The search for an unconventional "new style", from a "rationalist" perspective, as a pure construction, was not rare in the Piedmont region, particularly for explicitly utilitarian programmes.

At the *Scuola di Applicazione per gli Ingegneri* in Turin, when Bongioannini was a student and showed especial interest in industrial engineering, the professor in architecture was Carlo Promis (Savorra 2017), who also held the equivalent chair at the city's Accademia. He was substituted for just one year (1870) by count Carlo Ceppi, and then he leaved the post to Giovanni Castellazzi (Richelmy 1872; Pugno 1959; Vitulo 1993[29]), who proclaimed himself a student of Promis'. Also decisive in the field of construction, particularly in road- and railway-building, was the influence of Giovanni Curioni, teacher in different positions, from 1863, and repeatedly indicated as the effective Department Head.

5 A Model for Municipalities

In his illustrated book on schools (Bongioannini 1879) Bongioannini adopted the same architectural language. The activities of the Museum of Teaching and Education included annual conferences and, during those of 1876 and 1878, the first issue "school desks and houses" was presented respectively by Girolamo Buonazia, professor of Applied Mathematics at Florence's *Istituto Tecnico* and *Accademia di Belle Arti*, and by Antonio Labriola, the Museum's own Director (Tauro 1903[30]). The Secretary General, Martino Speciale, a lawyer and long-time member of Parliament—called upon to ensure the continuity of a ministry in which, due to the frequent government crises, Francesco Paolo Perez and De Santis succeeded Coppino—responded to the dedication to the Minister, a generic "His Excellency": he requested to provide two hundred copies to the Ministry, so that the book "may be made known to the provincial school authorities". He referred to an old eighteenth-century custom according to which the dedication corresponded to a purchase of copies, here pre-arranged within the same offices.

The types of buildings for primary schools (Fig. 1)—which, "through the explanation given of them last year at one of the School Museum's Teaching Conferences … gave rise to a decision … that the Municipalities should be informed of them …"—are delineated with the help of the *quadrillage* of Durand

[28]p. 76 (and photos nn. 127–128, p. 84).

[29]p. 63.

[30]p. 13.

who, in turn, loved the regularity of Turin (Szambien 1984[31]). Another reference to the teaching of the *École Polytechnique* is the division of buildings into modular elements, in this case the classroom with the vestibule. These, in turn, may be broken down into single square modules. With their porticoes in place of corridors, which were reserved—although it is not known how—for the harshest climates, sequences of columns and over-abundant distribution, arranged according to the same modularity, these school buildings were presented as a model of "rational composition" which would also give rise to "rational constructions". Composition and construction were originated by the function, in a programme combining traditional reflections on the choice of area with the most binding quantitative, regulatory and sanitary specifications of compulsory schooling. This approach involved "seeking, in all their essence, the most important monumental buildings, attempting to understand the ideas that shaped them … repeating, on each of them, all the workings of the mind that created them" (Bongioannini 1879[32]). The principles of Viollet-le-Duc expressed the great—and contradictory—desire of that century, namely progress freely built on the nation's past and individuality. Referring to these, Bongioannini sought to provide school buildings with the necessary link to history as well as the essential didactic dimension. Rational construction aimed to emphasise the regular stone or brick load-bearing structure, clearly distinguished from "infills", like Gothic and even "Lombardesque buildings" (Bongioannini 1879[33]), a further, generic, *filius temporis* homage to the myth of the rationality of medieval construction. Of course, "the decorative parts are not achieved through superfetation but by embellishment and ornamentation of the constructional parts, while, for the classrooms, successive… openings in the form of a continuous window … are envisaged … divided only by pillars or small columns" (Bongioannini 1879[34]). The text, often confused and redundant, particularly in its historical *excursus*, conceived—it would seem—in homage to the principle "*repetita iuvant*", shows via which humble, involuntary routes, that Kaufmann would never have imagined, it is possible to slowly advance on the path "von Ledoux bis Le Corbusier". It is therefore advisable to focus on building health and trace another stage in the development of standards. The seventy student classrooms stated in the Regio Decreto no. 4684 were limited to fifty students, each of whom were entitled to a volume, corresponding to the height, of five or six metres, depending on the clemency of the climate; 13–15 mq were also added for the teacher. The air had to be renewed two or three times per hour, and the glazed surfaces—1.5 m above the ground, in order to render them inaccessible to children—had to be a minimum of 1/40 or 1/30 of the volume, i.e. 1/5–1/3 of the floor area. While uncertain and variable, standards were taking a direction. Heating was provided by a stove located in the vestibule, on the wall towards the

[31]p. 34 (footnote no. 17).

[32]p. 8.

[33]p. 34.

[34]p. 24.

classroom (Fig. 2). The fireplace was fed by "stale" air, which entered via a vent opened at the level of the classroom floor and exited as smoke via the chimney flue. Instead, fresh air entered via the grates to the sides of the stove, flowed around the stove and in a large flue within the wall thickness, surrounding the iron smoke flue, and warmed up exited in the classroom, just below the ceiling. References to use of this equipment as a summer ventilation system, activated by a flame of obviously reduced intensity, perhaps limited to a gas or acetylene burner, are unconvincing. The teacher played a key role, being able to open the window transoms and operate the external tiltable shutters ("cord shutters") or "Chinese-style shades", consisting of thin, fixed, wooden battens, and the internal blinds in order to protect preferably south-facing rooms from solar radiation. Bongioannini made little use of the notions exhibited ten years earlier; indeed, these guidelines and tools for climate control appear incoherent and, above all, vague, as though the interest of users and the ability of the municipalities to apply them were very scarce.

6 Camillo Boito and the Schools of Padua at the Reggia Carrarese

The primary school designed by Camillo Boito for the city of Padua briefly pre-empted the first two stages of the regulations about the school building. This was, perhaps, no coincidence: he succeeded in skilfully presenting his work as an example in the official spheres and broadcasting it to the public of technicians and administrators, and was able to obtain resources far superior to those granted to other architects. The documentation demonstrates that his buildings were carefully considered and, to varying extents, imitated. The reports and articles illustrating them and the publications citing them also offer a broad picture of the European literature of the subject and therefore of the references known to the technicians and educated Italian public of the period, which, indeed, were promptly reflected in constructions. In fact, they set the standards for more complex school buildings, also in terms of well-being and installations (Figs. 3, 4 and 5)

In Padua, Boito won the assignment on 6th March 1877, and the building was inaugurated at the end of September 1880 (Serena 2000; Vettori Ardinghi 2012). The Minister Coppino had presented his bill on compulsory schooling to the Chamber on 9th February (Talamo 1983), and the law was approved on 6th July of the same year. In this singular interweaving of dates, the Padua initiative would seem to suggest tangible political support for a Minister, Coppino, who was in a broader sense, man of culture, active not only in the education problems but also in the issues of monument care. He proposed a bill on the subject to the Senate on 3rd February 1877 (Nicolini and Sicoli 1978[35]). In 1888, his subsequent proposal, already approved by the Chamber, was rejected by the Senate by secret ballot (Bencivenni et al. 1992). On that occasion, again, Boito assured his support to

[35]Chronological table of parliamentary debate at pp. 64–68.

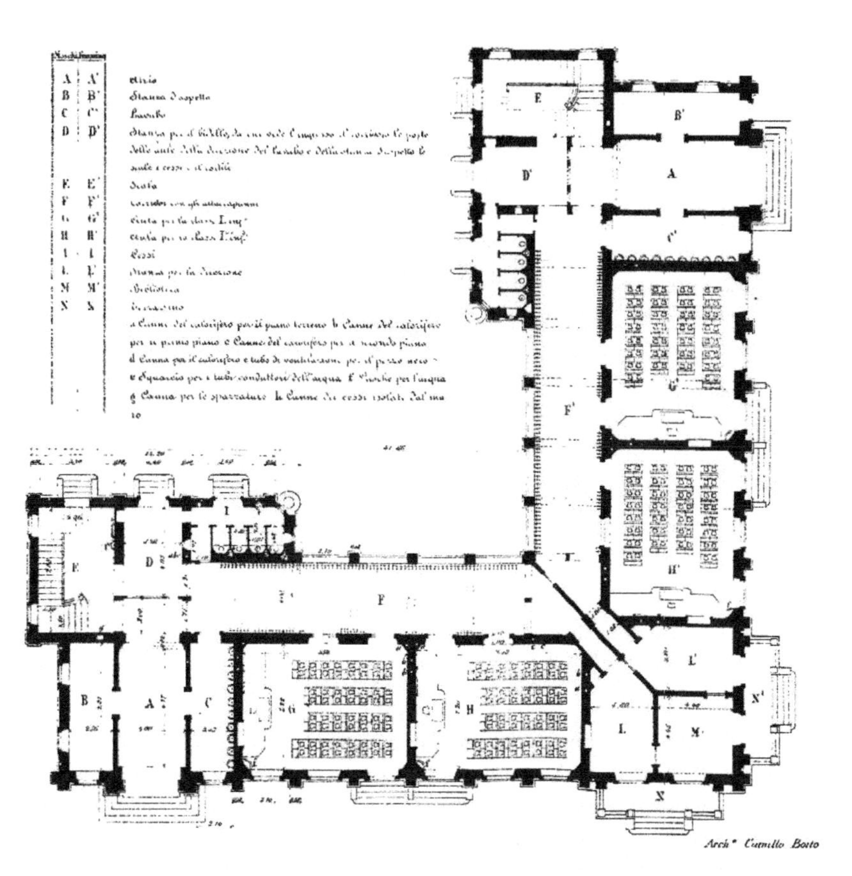

Fig. 3 Primary school at the "Reggia Carrarese" (Padua, It), groundfloor plan, Vittanovich (1885), vol. XXXIII, tav. 9

Coppino (Zucconi 1997[36]). In his report of 29th August 1877, Boito repeatedly cited the work of the Commission established by Bonghi, particularly when he claimed not to adhere to its guidelines. His building comprised three storeys, in contrast to the prescribed two, and his orientation draws the greatest possible benefit from the area, as is essential when situated within the urban fabric of an historic city. The "large, comfortable, well-lit […] 5-m-high classrooms housed less than 70 students for the first and second classes and approximately 40 for the third and fourth class. The windows were separated by "very narrow piers in order to avoid

[36]pp. 265–266, no. 58.

Fig. 4 Primary school at the "Reggia Carrarese" (Padua, It), cross section and internal front, Vittanovich (1885), vol. XXXIII, tav. 9

shadows", with sills 1.5 m above the ground, and their surface area was between a quarter and a fifth of that of the rooms, equal to 70 m^2. The plan for heating and ventilation was more complex: the Commission "elected by the Council", which followed the project from its origins, did not always agree with Boito's proposals. He had to renounce the vents at the base of the walls, under the windows, for "eradicating the impure air of the lower layers". The squares, clearly visible in the design drawings, disappeared in the subsequent engravings. Ventilation was provided by the window transoms, which were quadripartite in order for the panels to fold into the wall thickness. Heating, to be "artificial, even on the harshest winter days", was opposed by the Commission, for "extremely serious reasons", namely the running costs which would be borne annually by the Municipality and, consequently, by the tax-paying citizens. Hostility towards mandatory schooling itself was rife among the influential figures of the Lombardia and Veneto regions, and the gaps in the Casati Law were an advantage not secondary to National Unity… With the exclusion of individual stoves and "hot water heaters, extremely costly to install and maintain", air heaters "which do not cost much and are easy to repair" were suitable for providing 12–13 °C, but 8 °C or 9 °C would have sufficed in the new school (Fig. 8).

Fig. 5 Primary school at the "Reggia Carrarese" (Padua, It), East front (1988)

Boito succeeded, at least, in enforcing the installation of "horizontal pipes between the floor and vaults, and vertical ones in the walls". While somewhat unclear, this phrase recalls a proposal by Rinaldo Ferrini—a colleague at the Politecnico—who contemplated, for schools, the circulation of hot air under the

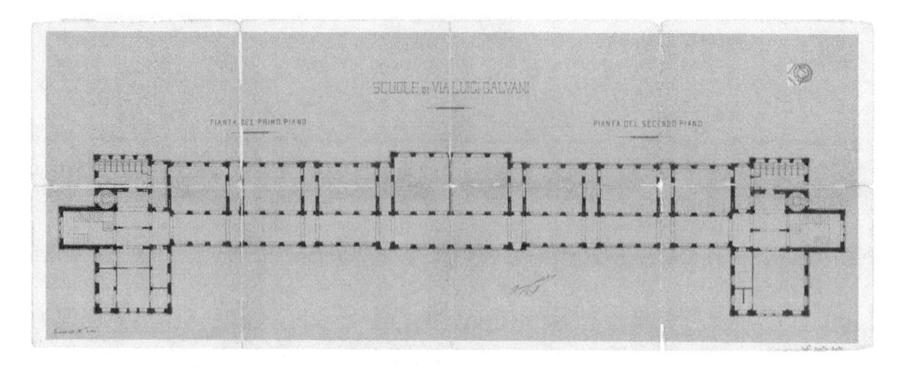

Fig. 6 Primary school in via Luigi Galvani (Milan, It), plan of the first and second floor, 1 February 1888, original scale 1:100. In: Archivio Civico Milano, Beni Comunali, Finanze, cart. 226

floor, having it rise in the front of the desks (Ferrini 1876[37]). In the end, on 29th November 1879—among the other alterations which significantly increased the estimated cost of 265,000 lire—a contract was entered into with Giuseppe Pollino for the construction and installation of two heaters at the "greatest possible reduction on the forecasted 5200 lire. The heater vent covers throughout the establishment must be made from brass". This was the most widespread solution at the time in dwellings, with vents at the base of the walls (Curioni 1873[38]; Sacchi 1874[39]). Among the drafters of the most diffuse manuals, Curioni, in particular, appears to have been a promoter of natural circulation, introducing clean air from below and removing it from above. Sacchi varied according to the circumstances, and appears to have preferred inverted circulation for very crowded environments. Ferrini also, opted for this solution: that appears contradictory. He was familiar with the work of Reid at the House of Parliament (Ferrini 1876[40]) and used the illustrations from his treatise in which natural circulation is increased using extractor chimneys (Manfredi 2013[41]). The authority of Morin and the French experience prevailed, in part due to the variety and breadth of testing of centralised systems.

At that time the standard (Ferrini 1876[42]; Sacchi 1874[43]) involved an internal temperature of 15 °C when 0 °C were detected outside. The French regulations of 1880 (Règlement 1881[44]) also set similar values (between 14 and 16 °C). The air

[37]p. 451.

[38]p. 236.

[39]p. 678.

[40]pp. 384, 441, figg. 101–104.

[41]pp. 208–209.

[42]p. 388.

[43]p. 656.

[44]p. 9.

Fig. 7 Primary school in via Luigi Galvani (Milan, It), plan of the first and second floor. Front detail on the street: the air vents viewable below the windows have been later optured. *Credits* Valisi (2014)

Fig. 8 A stove (patented by Louis Frédéric Staib) is described in the printed catalog of the Edoardo Lehmann company (1888)

flowing out of the vents should, once again according to Ferrini, range from 35 and 40 °C, at a velocity of 0.4–0.7 m/s: these were the limits typical of natural circulation. However, the temperatures actually achieved remained uncertain: Cantalupi indicated temperatures of 16–18 °C to be achieved in meeting rooms (Cantalupi 1862[45]), while the tables for dimensioning of heaters cited by Sacchi refer to temperatures of around 13 °C (Sacchi 1874[46]). The variations are probably due to the presence of people and to gas lighting.

The dimensions adopted in the Reggia Carrarese were simply repeated by Bongioannini in 1879, while, in Boito's work, they were translated into a certain architectural concept in which intentions are rendered concrete in design with the maximum coherence. This explains the success of the school in Padua, the visit from the Mayor of Venice on 6th March 1881 and the sending, in 1882, of eight photographs to the municipalities of Rome, Naples, Turin, Milan, Florence, Verona and Venice (Vettori Ardinghi 2012[47]), the Museum of Teaching and Education of Rome and the Museum of Education newly founded at the University of Palermo.

In 1885, Pietro Vittanovich, the Padua superintendent of schools, and a teacher at the local secondary school illustrated the building in the *Giornale dell'Ingegnere* (journal and organ of the *Collegio*, the Milanese Association of Engineers) reproducing word-for-word, without quoting it, the very extensive steps of Boito's handwritten report (Vittanovich 1885). The article was then re-printed as an extract, with drawings corresponding to the constructed building. It cited numerous publications from the rest of Europe and the United States. Some of the books were available in Milan, kept in the library of the Politecnico, and the majority in Rome, at the Museum established by Bonghi, responsible for gathering documentation on school buildings.

The keen interest of architects and engineers is clearly visible both in the proliferation of local pamphlets, often difficult to obtain, and in the summary—a low-cost publication—by Giovanni Sacheri: the international scene is reconstructed, above all, by the materials exhibited or illustrated at the World Fairs of Vienna (1873) and Paris (1878) which explain "exotic" references, such as Norwegian schools (Sacheri 1883).

7 The Schools of via Galvani, Milan

The importance that Boito attributed to school architecture is confirmed by his design for the 24-classroom complex of the industrial district to the north of the city, a context very different to the historic centre of Padua, yet equally

[45]p. 356.
[46]pp. 625, 627.
[47]p. 20.

high-profile.[48] The Mayor, Gaetano Negri, awarded him the assignment on 7th April 1887, and the design was delivered on 6th November of the same year. On 13th December 1887, the City Council requested a reduction on the estimated cost of 610,000 lire. On 25th January 1888, Boito presented an alteration reducing the costs to 496,000 lire, which was approved by the City Council on 4th February. The central pavilion was re-sized: the first design planned a two-storey gymnasium together with a large drawing studio on the third floor, flanked on both sides by classrooms for manual work. Only the ground-floor gym remained. On final balance, the costs rose to 583,000 lire, that is 24,000 lire per classroom. The councillor who complained of costs tripling those of other schools designed by the Technical Office was, in any case, in error. Numerous opinions underlined—in agreement with the architect's report—the value of the building as a model and, indeed, the need to further raise the standard of school premises (Figs. 6 and 7).

The classrooms, measuring 61.5 mq, were designed for 48 places. The surface area of the three windows, 1.4 m above the floor, totalling 15.3 mq, equalled a quarter of the floor plan. The wall on the corridor side was—unlike in Padua—opened up by windows corresponding to the external ones, and all of them were fitted with transoms for ventilation. Under the central windows, as many vents were created in the base of the external walls and the spine wall, measuring 1.2 × 0.2 m, then openable and now blocked but easily recognisable due to their stone cornices, to improve ventilation. Bongioannini (Bongioannini 1870), followed by Ferrini, had already recommended air intake vents at the base of the walls and warm air outlets near to the ceilings, a mixing system closer to the French studies than to the English experiences, in which ventilation air was drawn in from below and out from above, thus using natural heating. However, extraction from below was, from a physical perspective, quite different from drawing in fresh air, as occurred in this case. In the lateral walls, channels were, in any case, provided in order to activate artificial ventilation, heating the air within the ducts using a gas burner to achieve an updraft.

However, heating, closely linked to ventilation, remained uncertain. Indeed, on 24th November 1887, the Health Commission recommended low-pressure steam heating, currently being installed at the premises of another school building, the *Società di Incoraggiamento Arti e Mestieri*. A long tradition was coming to an end, and technological progress was now rendering these more complex systems reliable.

The uncertainty of the report was pure opportunism, because the decision, partly due to cost limits, had already been made and the masonry ducts are clearly visible in the design drawings, attached to the ruling of 4th February 1888. However, only upon request by Boito himself, on 13th May 1889, did the procedure for selection and assignment by direct agreement begin. Five firms—Zanna, Guzzi & Ravizza,

[48]All the documents concerning the school of via Galvani—if here are not different indications—are located in in Archivio Civico del Comune di Milano (hereinafter ACMi), Beni Comunali, Finanze, cart. 226–227.

Mussi, Lehmann, and Besana & Carloni—had submitted proposals, and a commission (consisting of architect Angelo Savoldi, Boito himself and, upon a Boito's proposal of his colleagues, Rinaldo Ferrini and Cesare Saldini) had limited the field to air ventilation and to three firms: Besana & Carloni, Guzzi&Ravizza, and Lehmann. The latter would be awarded the contract in accordance with a prefectural authorisation of 29th August.

8 The Companies' World?

The procedure followed makes it possible to reconstruct the state of the art in the sector, and—by comparison with conserved documents regarding other, similar installations—the ongoing developments. Renouncement of "maximum high-pressure" steam was due to its installation cost, correctly valued as double that of an air system, its burdensome maintenance and the need for specialist personnel to manage it.

Air heaters, particularly in Milan where an early example dates back to the mid-eighteenth century (Forni 1997), was already traditional; indeed, advanced versions were also available which heated a lot of air to a relatively reduced temperature, while the "old-fashioned" models still offered little air at high temperature. These were not new issues but, rather, reflected the literature of ten or even twenty years before. Air heaters were merely large "circulation" stoves and also offered advantages in terms of scale. As the size of a building increases, it is advisable to centralise management in dedicated areas. Otherwise, it becomes necessary to transport fuel to higher floors, along corridors, at the expense of cleaning. With regard to their performance and actual savings, opinions differ, for example, between the young Bongioannini and Ferrini. Furthermore, much of the construction was in masonry, and the technology focused on the stove and on heating of the air rather than on its circulation. This was the focus of research in the '60 s, after the improvements introduced by Meissner in the 1820s (Forni 1997). Indeed, it would be tempting to say that research had ground to a halt: the manufacturers who introduced the most significant alterations, Zanna and Monti, were local, the latter continuing a firm established by Duke Antonio Litta in 1857. The former (Manfredi 2013[49]; Manfredi 2017[50]) probably relied on the Viennese experiences of the *Vormärz* and had established himself in Turin in 1852. The hot smoke of the inverted circulation stove was contained in two kinds of superimposed chambers made from metal sheet. Through these passed pipes channelling the outside air to which the heat was transmitted.

Duke Litta had purchased a French patent (Cantalupi 1862[51]) filed back in 1839 by the engineer Bernard Chaussenot Jr. and, in this case, the stove featured a

[49]pp. 189–190.

[50]p. 53.

[51]p. 493.

hemispherical upper space for collection of the rising smoke, which then passed through a dense series of pipes and descended into a lower, equal-sized space beneath the stove. Each of these firms had headquarters in both Milan and Turin. In the first large, post-unification school building—that of corso di Porta Romana, at the corner of via Rugabella, Milan—the designer (the engineer Agostino Nazari) proposed "Litta system" heaters,[52] while, for the other large complex on via Santo Spirito—via Spiga—via Borgospesso, he repeatedly used Zanna ones, from 1870 to the 1880s.[53] Assignment was awarded through restricted invitations to tender. Alongside the best-known and most technically reliable firm, there was a world of "stove fitters" with shops and warehouses in the old centre of the city; small businesses that assembled pipes and cast-iron sheets or metal plates, perhaps supplied, in part, by larger companies. In other cases, such as Guzzi & Ravizza, a professional who designed the installation competed by probably coordinating workshops that performed the work. This was an initial step towards the birth of a true business. A summary overview of the most innovative companies was provided by Baseggio in the magazine *L'Edilizia Moderna* of 1892 (Baseggio 1892[54]). These were predominantly concentrated between Milan and Turin, but their operating radius, after the unification, was nationwide, as demonstrated by the reference pamphlets printed several times by Litta—G. B. Monti, the Lehmann catalogues and the references included in the Besana & Carloni tenders.

Edoardo Lehmann,[55] who was selected for the schools of via Galvani, proposed a patent filed in Geneva, by Louis Frédéric Staib (1812–1866), a solution also dating back to the 1850s and subsequently perfected (de Candolle 1867[56]; Wartmann 1873[57]). Bongioannini illustrated it and Ferrini described it using the name of Staib's successors, Weibel&Briquet (Weibel 2006). In a masonry parallelepiped, into which outside air flowed, the stove and smoke evacuation pipes were contained within a smaller, cast iron, concertina-type parallelepiped. This increased the ventilation surface area and transmitted the heat at a lower temperature to a larger quantity of air (Fig. 8).

The commissioners had no doubts despite the fact that the cost exceeded that proposed by Besana & Carloni by 3000 lire, even when, having requested a reduction to 12,500 lire, they had to agree on 13,000 lire. In their opinion, the quality-price ratio was still decidedly superior. *In tempore non suspecto*, the

---

[52]ACMi, Beni Comunali, Finanze, cart. 209, "Perizia del 28/3/1864".

[53]ACMi, Beni Comunali, Finanze, cart. 220, fasc. 9, "Prospetto riassuntivo delle quantità delle opere occorrenti per l'esecuzione di parte del progettato fabbricato… di Via Santa Spirito, 23 aprile 1870" and "Scuole in via Borgospesso—costruzione del calorifero, 18-28 luglio 1878".

[54]pp. 4–6.

[55]The firm was operating till 1906, when was replaced by the firm Haeberlin Gerra & C., (1907–1913), in Archivio Storico della Camera di Commercio di Milano, Monza, Brianza e Lodi, Archivio Ditte. http://www3.milomb.camcom.it/index.phtml?pagina=form&nome=ARCHIVIO_T_Ditte&explode=10.05&azione=UPD&Id_Ditte=27423, consulted on 28/12/2017.

[56]Obituary, pp. 288–290.

[57]pp. 68–69.

technical literature had already adopted a position, by them or by colleagues they cited with the due esteem. In proof of this, Boito entrusted the installations of his last great work, the Casa di Riposo per Musicisti (a retirement home for opera singers and musicians in Milan), to Lehmann.[58]

Of Swiss origin, and a francophone despite his surname, as may be deduced from the rare and minor errors in his clear and correct Italian, Lehmann had established himself in Milan in 1879 as holder of a valued patent. In 1886, he completed his workshop, a block between the present via Lazzaretto, via Casati and viale Tunisia, at that time next to the railway tracks forecourt of the Central Station.

His project report reveals how the literature was transformed into high-quality practice, and how, during this practice, the resolution of certain problems developed. The building would have required six heaters, albeit of different powers, confirming the regulatory limit—15 m—for horizontal air paths. The ducts issued the hot air into the classrooms at 2.5 m above the floor and were constructed from hollow bricks which served as insulation. Additional hollow bricks doubled the solid brick wall of the boiler. Lehmann showed scepticism with regard to the ventilation shafts, which were too numerous—one per classroom—not to impair the function of the spine walls. The extraction activated by gas flame, when the boilers were switched off, would not, in practice, be used, and opening of the windows, which it was impossible to prevent, would, in any case, have undermined its function. As a last resort, it was more advisable to input forced air using a fan operated by gas motor which served to circulate the water, which, pending waterworks with adequate pressure, was, in fact, purchased. *Mutatis mutandis,* the famous alternative of the Hôpital Lariboisière was thus revived (Manfredi 2013[59]).

9 Towards New Climate Control Systems in School Buildings

A useful frame of reference are the conditions desired, more than imposed, by Regio Decreto no. 5808 of 11th November 1888 for school buildings funded by loans partially borne by the State. These regulations finally established a maximum number of fifty students per classroom, recommended separate areas for the changing rooms and a transparent surface equal to a quarter of the floor area, and permitted three storeys in urban centres. With regard to ventilation, transoms were required above the windows. Using heating, temperatures of 14–16 °C (the French standard) should be achieved by inputting air at a maximum of 60 °C (well above the 45 °C limit admissible according to Ferrini) and at a height of three metres. In addition, “the stale air output vents should be set at floor level”. This was a

[58]Archivio della Casa di Riposo per Musicisti—Fondazione Giuseppe Verdi, Fondo economato, b.218; cfr. http://lombardiarchivi.archimista.it/fonds/35045 cons. on 30/12/2017.

[59]Overall pp. 135–136 with bibliography for preceding texts.

fragmented, confused and limiting set of instructions compared to the "best practices" of the period, and insufficiently applicable to the schools of smaller centres, the most numerous and necessary.

Another frame of reference is provided by other Milanese schools. Between 1886 and 1889, the director of the via Galvani works, Angelo Savoldi, had built the schools of via Felice Casati, an initial attempt to propose more advanced school building standards.[60] Few documents have been conserved, and it is necessary to refer to the numerous publications,[61] testifying to their success and value as a model. The cost—approximately 454,000 lire—was just under 18,000 lire per classroom, 25% less than Boito's. The transparent surface had been reduced to a fifth of the area of a classroom, and the width of the corridors had been decreased to three metres. Climate control, both active and passive, was more complex: the windows were screened by high, narrow, four-part blinds which could slide in two different directions and back into the wall. The parapets consisted of two slender walls which enclosed a cavity. To the exterior, there was a series of terracotta rose windows. To the interior of these, there was a slit the length of the splay, enclosed by a metal frame with two sliding, perforated plates for blocking or opening the flow of air "like those of railway carriage ventilation doors". Four heaters—one for each wing of the building—were also present, manufactured by Guzzi & Ravizza. The hot air rose in the spine walls and was input into the classrooms at 1.8 m from the floor. In the skirting, on the other short side, were positioned the vents of the ventilation ducts, which rose up to the roof where they joined the chimney flues of the heater boilers, and the hottest smoke activated the outflow. In summer, at the point of the extraction chimney, the customary solution of the gas flame would have been adopted (Figs. 9 and 10).

The air heater, entrusted to Lehmann, appears in one singular case: the Realdo Colombo school in Cremona, whose designer, the city engineer Pietro Ghisotti, personally visited the schools of via Galvani and via Casati and consulted the documentation pertaining they (Landi 2017[62]).

The via Galvani school was opened in autumn 1890, while the rapid population growth necessitated increasingly large spaces. During the second half of 1891, a City Council commission established the relative financial plan for three hundred classrooms over the decade at an average cost of 12,000 lire per classroom, half the amount Boito had calculated. This drastic reduction dictated a further simplification of the already minimal decorations. The changing rooms were renounced, the corridors were reduced to 3 m in width, the transparent surface decreased to a fifth or a sixth of the area of the classrooms, and the height of these dropped to 4.7 m;

[60]The contract with the master builders Maggioni e Mazzai was underwritten on October 20, 1886 and the provisional trial was signed by the architect Giovanni Ceruti, on January 24, 1889 (ACMi, Beni Comunali, Finanze, cart. 223, fasc. 1).

[61]Its most complete description perhaps in Mina (1892).

[62]pp. 162–164.

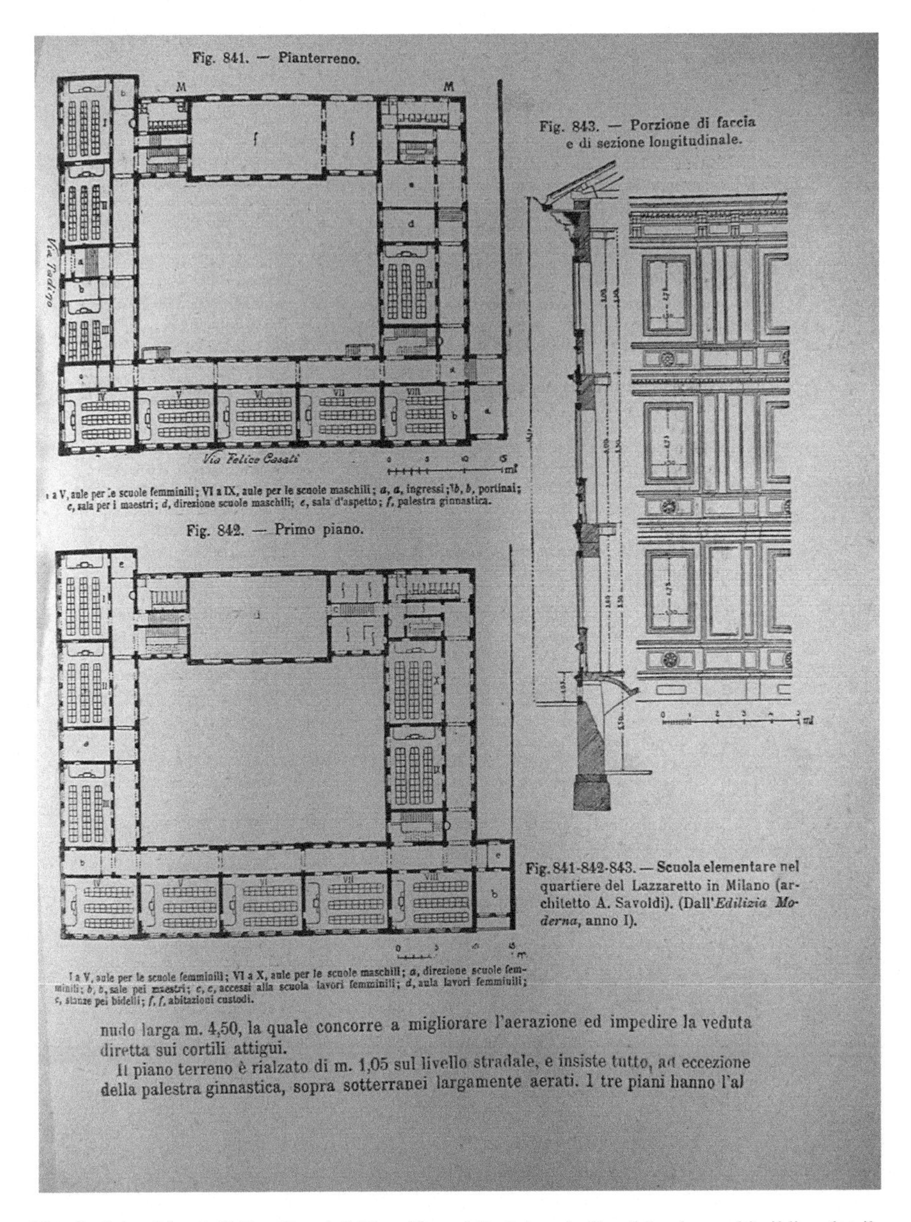
Fig. 841. — Pianterreno.

I a V, aule per le scuole femminili; VI a IX, aule per le scuole maschili; *a*, *a*, ingressi; *b*, *b*, portinai; *c*, sala per i maestri; *d*, direzione scuole maschili; *e*, sala d'aspetto; *f*, palestra ginnastica.

Fig. 842. — Primo piano.

I a V, aule per le scuole femminili; VI a X, aule per le scuole maschili; *a*, direzione scuole femminili; *b*, *b*, sale pei maestri; *c*, *c*, accessi alla scuola lavori femminili; *d*, aula lavori femminili; *e*, stanze pei bidelli; *f*, *f*, abitazioni custodi.

Fig. 843. — Porzione di faccia e di sezione longitudinale.

Fig. 841-842-843. — Scuola elementare nel quartiere del Lazzaretto in Milano (architetto A. Savoldi). (Dall'*Edilizia Moderna*, anno I).

nudo larga m. 4,50, la quale concorre a migliorare l'aerazione ed impedire la veduta diretta sui cortili attigui.

Il piano terreno è rialzato di m. 1,05 sul livello stradale, e insiste tutto, ad eccezione della palestra ginnastica, sopra sotterranei largamente aerati. I tre piani hanno l'al

Fig. 9 School in via Felice Casati (Milan, It), architect Angelo Savoldi, plan and building details: *L'edilizia Moderna*, 1892 fasc. V, p. 5

the 30 cm floor permitted a storey height of five metres, a ploy in order to comply with Regio Decreto no. 5808. As far as heating was concerned, the air heater was once again adopted.

Fig. 10 School in via Felice Casati (Milan, It), architect Angelo Savoldi, front on via Tadino, in: *L'edilizia Moderna*, 1892 fasc. V, tav. XXIII

The extent to which the spending plans were adhered to remains to be examined. Certainly, air heaters were, in more than one case, replaced by steam heaters. In 1893, the company Piazza & Zippermayr,[63] for example, manufactured central heating apparatus costing 26,000 lire for the Via Ariberto School; a single boiler which, however, served no less than 45 classrooms. This was a further step towards the "low-pressure water heating" which would triumph in the following century.

References

Baseggio N (1892) Il riscaldamento a vapore negli ambienti abitati. In *"L'edilizia moderna"*, fasc. VIII, pp 6–8 e fasc. IX pp 4–6

Bencivenni M, Dalla Negra R, Grifoni P (1992) Il decollo e la riforma del servizio di tutela dei monumenti in Italia, 1880–1915, vol II, Ministero per i beni culturali e ambientali, Soprintendenza per i beni ambientali e architettonici per le province di Firenze e Pistoia, Firenze

Boito C (1873) Architettura. Relazione di Camillo Boito. In: Relazioni dei giurati italiani sulla Esposizione universale di Vienna del 1873, vol I, fasc 5, Milano, pp 17–29

Boito C (1916) a cura del Comitato per le onoranze alla sua memoria. Allegretti, Milano

Bonghi R (1873) Relazione di Ruggiero Bonghi sulla educazione, istruzione, coltura quale era rappresentata all'esposizione universale di Vienna nel giugno 1873. In: Relazioni dei giurati italiani sulla Esposizione universale di Vienna del 1873, vol I, fasc 1, Regia Stamperia, Milano

Bongioannini F (1870) Riscaldamento e ventilazione dei luoghi abitati: norme pratiche e teoriche per l'impianto dei migliori sistemi: dissertazione presentata da Francesco Bongioannini per ottenere il diploma di ingegnere civile, Foa, Torino

Bongioannini F (1879) Gli edifizi per le scuole primarie [S. l., s. n.], tip. Artero & C., Roma

Bongioannini F (1898) Appunti per la conferenza sul tema Gli edifizi scolastici (Congresso pedagogico Nazionale in Torino, settembre 1898), Grato Scioldo Editore (tip. lit. Camilla e Bertolero di N. Bertolero), Torino

Bongioannini F (1909) Per l'ordinamento del servizio tecnico della pubblica istruzione. Paravia, Torino

Boswell Reid J (1844) Illustration of theory and practice of ventilation. Longman Brown Green & Longmans, London

Brizio E (1872) Scavi nel Foro Romano. In: Bullettino dell'Instituto di corrispondenza archeologica. settembre–ottobre 1872, pp 225–236

Cantalupi A (1862) Istruzioni pratiche elementari sull'arte di costruire le fabbriche civili. Salvi, Milano

Colombo E (2005) Come si governava Milano. Politiche pubbliche nel secondo Ottocento, Franco Angeli, Milano

Cossetto M (1997) Il Museo della Scuola – Schulmuseum della città di Bolzano, i Musei pedagogici dell'Ottocento. In: *Schulmuseum - Museo della Scuola*. Selbstverlag, Bozen, pp 1–4

Cronaca dell'istruzione (1896) L'Osservatore Scolastico, a. XXXII, 10.10.1896 p 12

Curioni G (1873) Costruzioni civili stradali e idrauliche. A. F. Negro, Torino

de Candolle A (1867) Discours. In: Bulletin de la classe d'agriculture de la société des arts de Genève, no. 31, 31 déc 1867, pp 287–292

[63]The contract was subscribed on May 9th, 1893 (ACMi, Beni Comunali, Finanze, cart. 216, fasc. 3).

De Fort E (1979) Storia della scuola elementare in Italia, vol I. Dalle origini all'età giolittiana, Feltrinelli, Milano
Ferrini R (1876) Tecnologia del calore. Hoepli, Milano
Ferrini G (1892) Tipi economici di scuole elementari pel Comune di Milano. Scuole di Via Giusti, Via Torricelli e Via Ariberto. In: *L'edilizia Moderna* 1892, fasc VII pp 4–6 and fasc VIII pp 4–6
Forni M (1997) Il palazzo Regio Ducale a metà Settecento. Considerazioni sulla residenza. Rassegna di Studi - Civiche Raccolte di Arti Applicate del Castello Sforzesco, Milano
Forni M (2017) La stufa alla moscovita a Milano. Applicazioni di un sistema di riscaldamento ad aria calda nei secoli XVIII e XIX, in Manfredi, pp 58–111
Grimoldi A, Landi AG (2014) Structural problems in Italian school buildings of the late nineteenth century: the school "Realdo Colombo" in Cremona. In Meli R, Peña F, Chávez M (eds) Proceedings of 9th international conference on structural analysis of historical constructions, Mexico City, 14–17 Oct 2014
Melis G e Tosatti G (1999) I tecnici delle Belle Arti nell'amministrazione italiana in Varni Melis, pp 183–206
Koller MF (1800) *Der Praktische Baubeamter*, Ignaz Alberti's Witwe, Wien, IInd ed
La Rosa N (2011) Francesco Bongioannini e la tutela monumentale nell'Italia di fine Ottocento. ESI, Napoli
Landi AG (2017) Dalla stufa al calorifero. Il progetto del confort a Cremona XVIII e XX secolo, in Manfredi, pp 144–177
Manfredi C (2013) La scoperta dell'acqua calda. Nascita e sviluppo del riscaldamento centrale, Maggioli ed., Santarcangelo di Romagna
Manfredi C (2017) La fortuna dei sistemi di riscaldamento dall'Europa all'Italia dell'Ottocento: I progetti nei riscontri documentari e nella pubblicistica. In: Manfredi C (ed) Architettura e impianti termici. Soluzioni per il clima interno in Europa fra XVIII e XIX secolo. Allemandi, Torino, pp 19–57
Meda J (2010) Musei della scuola e dell'educazione. Ipotesi progettuale per una sistematizzazione delle iniziative di raccolta, conservazione e valorizzazione dei beni culturali delle scuole. In: History of education & children's literature, vol V, no 2, pp 489–501
Melis G (1996) Storia dell'Amministrazione Italiana (1861–1893), Il Mulino, Bologna
Miano G (1987) Giulio De Angelis. In: Dizionario Biografico degli Italiani, vol 33
Mina C (1892) Edificio per scuole elementari in Milano - Quartiere del Lazzaretto. Arch A. Savoldi, in *L'edilizia Moderna*, fasc V, pp 5–7
Movimento di Provveditori (1908) in "I diritti della Scuola", p 531
Nicolini P, Sicoli S (1978) Verso una gestione dei beni culturali come servizio pubblico: attività legislativa e dibattito culturale dallo Stato unitario alle regioni (1860–1977). Garzanti, Milano
Pisani Dossi CA (1884) I mattoidi al primo concorso pel monumento in Roma a Vittorio Emanuele II. Sommaruga, Roma
Portoghesi P (1968) L'eclettismo a Roma 1870–1922, De Luca, Roma s.d.
Pugno GM (1959) Storia del Politecnico di Torino, Stamperia artistica, Torino
Règlement pour la construction et l'ameublement des maisons d'école,extrait de la Revue Pédagogique, Delagrave, Paris (1881)
Richelmy P (1872) Intorno alla Scuola di applicazione per gl'ingegneri, fondata in Torino nel 1860. Cenni Storici e statistici, Fodratti, Torino
Sacchi A (1874) Le abitazioni. Hoepli, Milano
Sacheri G (1883) Dei migliori tipi di fabbricati per le scuole comunali, Tip. e lit. Camilla e Bertolero, Torino Camilla e Bertolero, 1883. (excerpt from: Le costruzioni moderne di tutte le nazioni alla Esposizione universale di Parigi del 1878, Torino)
Sanzo A (2012a) Studi su Antonio Labriola e il Museo d'istruzione e di educazione. Edizioni Nuova Cultura, Roma
Sanzo A (2012b) L'opera pedagogico-museale di Antonio Labriola. Carte d'archivio e prospettive euristiche, Edizioni Nuova Cultura, Roma

Sanzo A (2017) Il «Giornale del Museo d'Istruzione e di Educazione». Politica editoriale e studi comparativi in educazione. In: Educazione: Giornale di pedagogia critica, vol VI, no 1, gennaio-giugno, pp 100–119
Savorra M (2017) Carlo Lorenzo Francesco Promis, in Dizionario Biografico degli Italiani, vol 86
Scoppola P (1971) Ruggero Bonghi. In: Dizionario Biografico degli Italiani vol 12
Serena T (2000) Scuole Elementari alla Reggia Carrarese. In: Boito C (ed) un'architettura per l'Italia Unita. Venezia, Marsilio, pp 98–102
Serra E (1987) Carlo Alberto Pisani Dossi diplomatico, Franco Angeli, Milano. Le Lettere, Roma
Szambien W (1984) JNL Durand. Picard, Paris
Talamo G (1983) Michele Coppino. In Dizionario Biografico degli Italiani, vol 28
Tauro G (1903) Della necessità di ricostruire in Italia un museo d'istruzione e di educazione. In: Bollettino dell'Associazione Pedagogica Nazionale, no 5, pp 197–227. (published as brochure: Torino, Paravia)
Varni A, Melis G (eds) (1999) Burocrazie non burocratiche: il lavoro dei tecnici nelle amministrazioni fra Otto e Novecento. Rosemberg e Sellier, Torino
Vettori Ardinghi A (2012) Costruzione e trasformazioni di un'opera di Camillo Boito: la scuola elementare alla Reggia Carrarese in Padova, degree thesis, Politecnico di Milano, aa. 2011–2012
Vittanovich P (1885) Le nuove scuole costruite in Padova alla Reggia Carrarese dall'architetto Camillo Boito. In: Il Politecnico - il Giornale dell'Ingegnere e dell'architetto civile e industriale, febbraio 1885, pp 81–92 (published as brochure: Minerva, Padova)
Vitulo C (1993) Riflesssioni sulla vita di Carlo Promis adi documenti della Biblioteca Reale di Torino. In Fasoli V, Vitulo C (eds) Carlo Promis professore di architettura civile agli esordi della Scuola Politecnica, Celid Torino, pp 47–75
Wartmann EF (1873) Notice historique sur les inventions et les perfectionnements faits à Genève dans le champ de l'industrie et celui de la médecine. Genève, Bâle, H.Georg Lyon
Weibel L (ed) (2006) Jules Weibel, un industriel au cœur de l'Europe: lettres à sa famille, 1857–1886. Editions d'en bas, Lausanne
Zucconi G (1997) L'invenzione del passato. Camillo Boito e l'architettura neomedioevale, Marsilio, Venezia, pp 176 and 67–69 and no 10
Zullo E, de Angelis G (2005) Architetto Progetto e tutela dei monumenti nell'Italia umbertina. Gangemi, Roma

Tradition and Science: The Evolution of Environmental Architecture in Britain from 16th to 19th Century

Dean Hawkes

Abstract This contribution traces the evolution of 'environmental architecture' in Britain in the four centuries from the end of the 16th century to the threshold of the 20th. That period witnessed profound developments in the applied sciences and in its later part in the technologies of environmental management in buildings. The essay proposes that these developments may be characterised as an encounter between reference to *tradition* as a source of architectural knowledge and the application of *science* and its cousin *technology* in the production of designs for buildings. The argument is illustrated with 'case studies' of six significant buildings.

1 Introduction: Definitions and Background

> *Tradition*, the transmission of customs or beliefs from generation to generation.
>
> *Science*, the intellectual and practical activity encompassing the systematic study of the structure and behaviour of the physical and natural world through observation and measurement.[1]

One of the principal functions of building is to provide shelter from the natural environment. In his seminal book, *The Architecture of the Well-tempered Environment* (Banham 1984), Reyner Banham shows how the act of building has enabled humankind to,

> … produce dryness in rainstorms, heat in winter, chill in summer, to enjoy acoustic and visual privacy …

The knowledge that has allowed this achievement is a matter of *tradition*, precisely as defined above. Throughout the world, in all cultures and climates, fundamental understanding of how to dispose of available resources in the service of

[1]*The New Oxford English Dictionary*, Oxford University Press, Oxford, 1998.

D. Hawkes (✉)
Darwin College, University of Cambridge, Cambridge, UK
e-mail: duh21@cam.ac.uk

C. Manfredi (ed.), *Addressing the Climate in Modern Age's Construction History*,
https://doi.org/10.1007/978-3-030-04465-7_6

shelter is transmitted from generation to generation (Rapaport 1969). This is the basis not only of vernacular building, but, it may be argued, plays a part in the design of what we may define as 'self-conscious' architecture.[2]

At the end of the 18th century the historic basis of the environmental function of architecture was transformed by the arrival of new tools that emerged from the applied science and technology of the Industrial Revolution. At first these were systems of centralized space heating and ventilation that were first introduced into industrial buildings and were quickly transferred to other building types. These were soon joined by developments in interior illumination, first by gas and then by electricity. By the end of the 19th century it was possible to conceive of buildings for all purposes in which the internal environment was substantially provided by artificial means. To paraphrase Siegfried Giedion, 'Mechanization *had* Taken Command' (Giedion 1948).

2 Hypothesis

The aim of this paper is to explore the evolving relationship between *tradition* and *science* as it may be found in the architecture of Britain between the 16th and 19th centuries. The discussion begins at the end of the 16th century with the works of Robert Smythson and proceeds to the 17th century, when Christopher Wren, one of the leading scientists of the day, became England's greatest architect. Next we consider the work of the English 'Palladians', who in the 18th century adapted the 16th century Italian architecture of Andrea Palladio to the culture and climate of England. Then we cover the 19th century through studies of three buildings, from the beginning, middle and end, by three of the most important architects of the time, John Soane, Charles Barry and Charles Rennie Mackintosh.

My hypothesis is that, over these four centuries, as society underwent remarkable changes; in politics, population growth, urbanization and the transformative developments of the industrial revolution; architecture, specifically in its environmental purposes, evolved a complex and creative synthesis between the environmental strategies of *tradition*, as these were described by Banham, and the potentiality of the understanding that came with the conscious application of *science* and soon of *technology*. It is proposed that this analysis both illuminates the historiography of architecture in this period and has implications for present day environmental design practice. This proposition was eloquently expressed by T. S Eliot.

> Tradition is a matter of … wide significance. It cannot be inherited, and, if you want it, it must be obtained by great labour. It involves, in the first place, the historical sense … the historical sense involves a perception not only of the pastness of the past, but of its presence. (Eliot 1920)

[2]The author has explored this question in previous essays Hawkes (1996).

3 Six Buildings

3.1 *Building One: Robert Smythson, Hardwick Hall (1590–1597)*

Robert Smythson (1536–1714) was a near contemporary of Andre Palladio (1508–1580).[3] In earlier studies I have shown how, despite their apparent stylistic differences, their buildings share a common response to the physical climates in which they were built; Smithson to the temperate conditions of England, Palladio to the warmer state of northern Italy (Hawkes 1996). Hardwick Hall (Fig. 1) is the greatest of Smythson's houses and most clearly shows his acute understanding of the climate. The exterior of the house presents a powerful bi-axial symmetry, with the long axis of the rectangular plan oriented almost exactly on the north-south cardinal point. The enormous windows led to the house being described as having, "more glass than wall." At a time when artificial light was rudimentary, depending on candles and oil lamps, these flooded the interior with daylight. Along the centre of the plan there is a thick masonry wall that contains 28 fireplaces that provided warmth to compensate for the heat lost through the windows. Within the formal exterior the internal planning is surprisingly free and this reveals the subtlety of its response to the climate. The principal rooms are all located at the south end of the plan, where they receive direct sunlight. This brings warmth as well as illumination. The principal occupant of the house, the Countess of Shrewsbury, was 70 years old when the house was completed. Her personal apartments were at the south end on the first floor, protected from contact with both ground and sky. This is environmentally the best place in the building. In comparison the apartments of the lower members of the household were at the less agreeable north end.

This brief analysis indicates the sophistication of the environmental design of the house.[4] This was achieved at a time when it was not possible to measure the elements of climate; temperature and barometric pressure.[5] These were matters of subjective experience and building design was thereby a matter of tradition rather than science.

3.2 *Building Two: Christopher Wren, Sheldonian Theatre, Oxford (1663–1669)*

Christopher Wren (1632–1723) was a scientist who became an architect. This fact may be shown to have a profound influence on his architecture, in particular in its

[3]For a comprehensive study of Smythson's works see Girouard (1983).

[4]A more extended analysis is at Hawkes (2012).

[5]See Middleton (1969) for an account of the development of climate measurement.

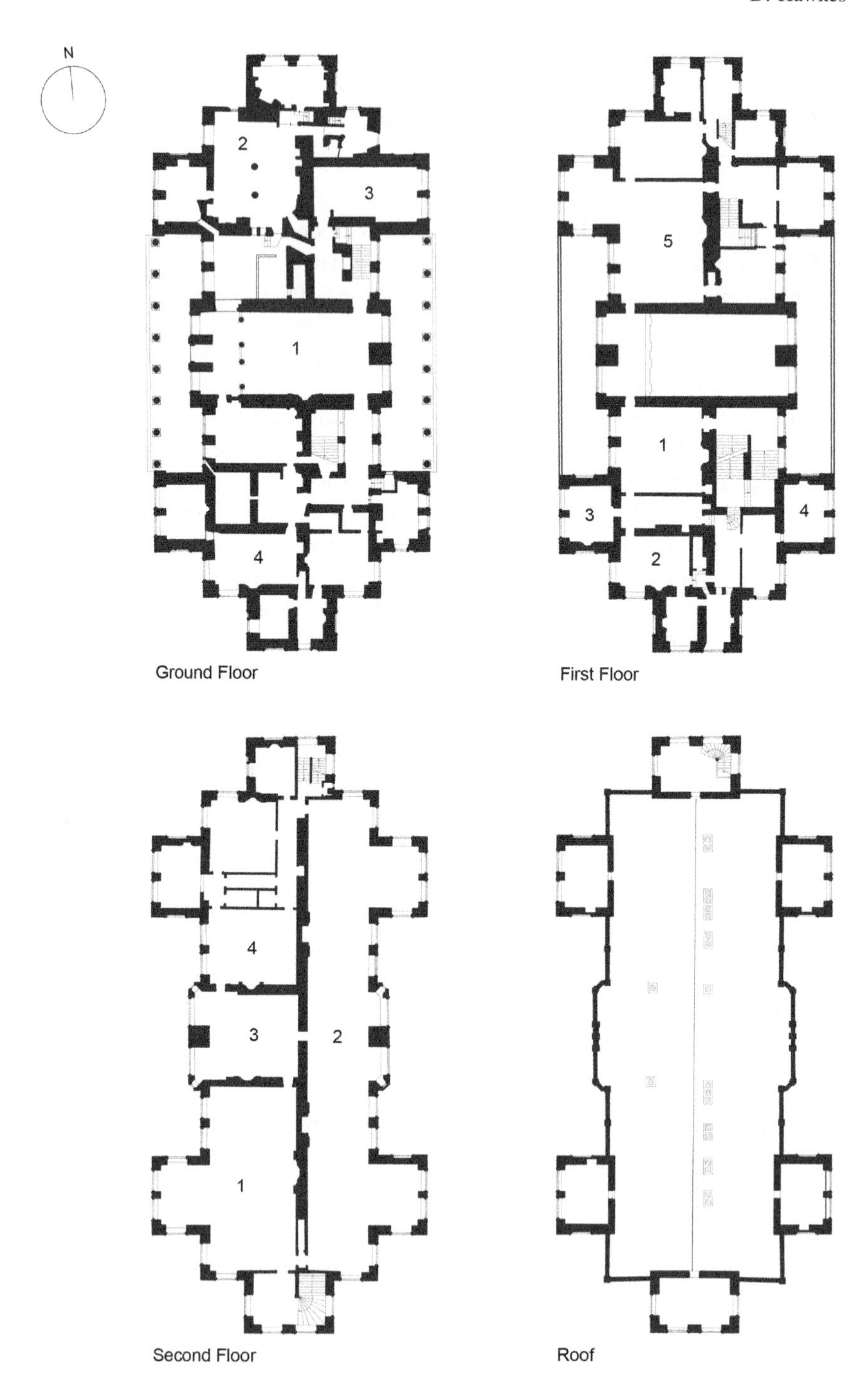

Fig. 1 Hardwick hall, plans

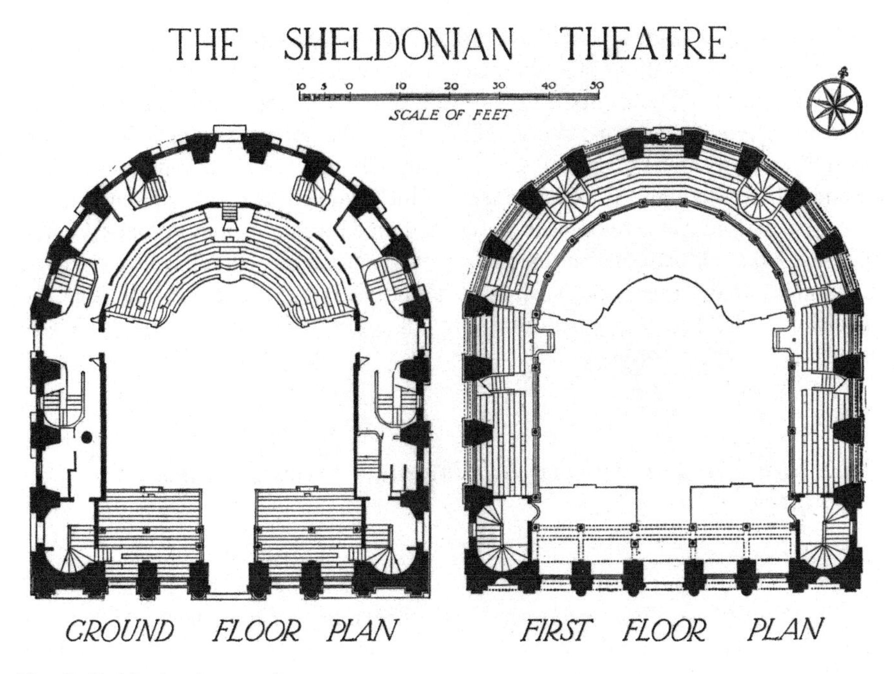

Fig. 2 Sheldonian theatre, plans

relation to climate.[6] Within the field of what we may define as 'classical' science, Wren made important contributions to mathematics and astronomy, but he was also interested in meteorology. In 1663 he presented a design for a 'Weather Clock', a portable meteorological station at a meeting of the Royal Society in London.[7] This was associated with an unrealized project to construct 'A History of the Seasons'[8] and measured and recorded air temperature, barometric pressure and rainfall.

In the year that Wren demonstrated his 'Weather Clock', he also exhibited his design for the Sheldonian Theatre at Oxford University at the Royal Society (Fig. 2) (Summerson 1632).[9] This building, which was built to accommodate the degree ceremonies of the university and also to serves as a place for musical performance, demonstrates how scientific understanding may be translated into architectural form. This is most clearly shown in the nature of its natural lighting. The amphitheatrical form of the building derives from the Roman Theatre of Marcellus, but unlike that precedent, the space is roofed. In the 17th century public

[6]This is discussed at Hawkes (2012, 2014).

[7]This illustrated in Summerson (1632).

[8]This is described in Wren (1750).

[9]The respective dates of the presentations were, Sheldonian Theatre, 29 April 1663, Weather Clock, 9 December 1663. This was the only occasion in his life-long membership of the Society that Wren presented an architectural project.

events took place during daylight hours and Wren's design fills the building with light from all orientations. On the exterior the windows constitute a quite small proportion of the envelope, unlike Hardwick Hall, but within the theatre there is an almost continuous band of large windows above the gallery seating, with a small band above the lower range of seats. In the language of 20th century building science, there is a high 'Sky Component' of the 'Daylight Factor'.[10] Wren worked, of course, before the development of our modern conventions, but I propose that in the Sheldonian Theatre and in the design of his numerous churches in the city of London and late masterworks, such as the library at Trinity College, Cambridge, the precision of the daylighting is informed by a scientific grasp that is entirely new in British architecture.

3.3 Building Three: Lord Burlington, Chiswick House, 1725–1729

Chiswick House in west London (Fig. 3) was designed by Richard, Lord Burlington (1694–1753) for his own use. The influence of Andrea Palladio is instantly apparent from the exterior, with its clear debt to the Villa Rotonda. Burlington was a leading member of the group that brought the ideas and image of 16th century Italian architecture to 18th century England, where it became the dominant influence on the design of country houses and other building types. Andrea Palladio's *I quattro libri dell'architettura* was published in English translation between 1717 and 1725.[11] Bringing Palladianism to England was not, however, an act of simple mimicry. The transition of place and time produced numerous variations of form and detail. The most significant of these were, perhaps, made in response to the very different climate.

The key to adapting to the different climate lies in the design of the windows. In Italy the primary question is to keep the house cool in the hot summer. In England the priority is for warmth in winter. The consequence is that the size of windows in Italy, relative to the volume of the rooms they serve, is smaller than that adopted in England. This is demonstrated if we compare Chiswick with La Rotunda. A curious fact in architecture is that theory sometimes follows practice. This is shown in the case of English Palladianism where, later in the 18th century, a number of significant treatises were published in which the question of window size was directly addressed. One of the most important of these is the *Lectures in Architecture* published in 1734–1736 by Morris (1734) and Hinchcliffe (2004). In

[10]See Hopkinson (1964) for a comprehensive account of the fundamentals of the science of daylighting.

[11]The text of the first English edition was translated by Nicholas Dubois with illustrations by Giacomo Leoni. Other editions were published by Edward Hoppus, 1736, and Isaac Ware, 1737.

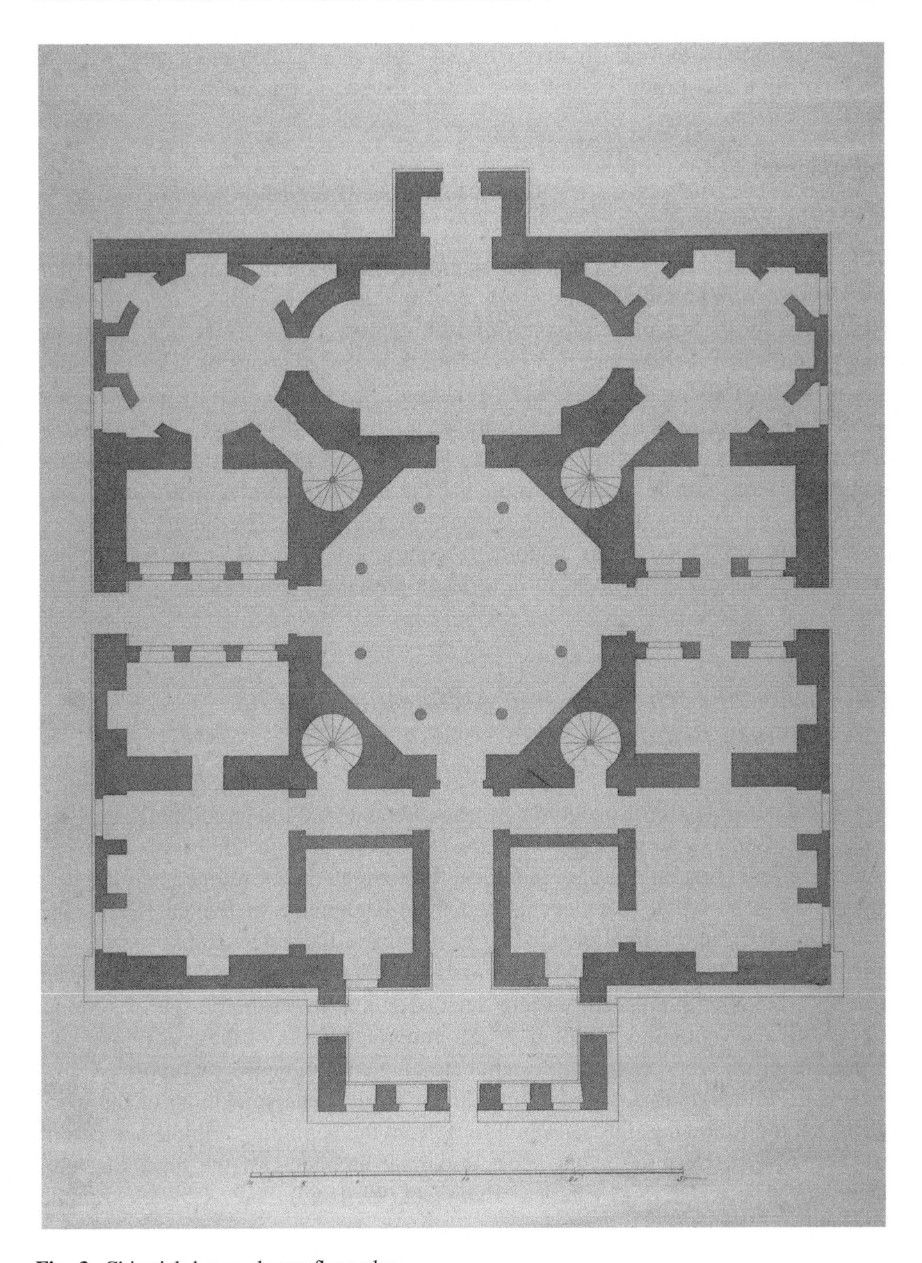

Fig. 3 Chiswick house, lower floor plan

this Morris adopts the Newtonian practice of quantifying and tabulating practical information, such as rules for establishing the dimensions of chimneys in relation to the rooms that they serve. On the question of window sizes in Lecture VII Morris provides a formula that clearly derived from Palladio's method in Chapter XXV in the First Book, but produces windows that are larger, in relation to the room, than the Italian prescription:

> multiply the Length and Breadth of the Room together, and that Product multiply by the Height, and the Square Root of that Sum will be the Area or superficial Content in Feet, etc. of Light requir'd.

In this Morris brings together the renaissance tradition of harmonic proportion that derives from Palladio with the new conception of tabulation of data that were established by the scientific progress of 18th century England. By this means we may propose that architecture had moved on to a new relationship with tradition.

Chiswick House anticipates these principles, with its carefully proportioned windows that amply light the interior under the dull English sky. The building is orientated at 45° from north. This brings Burlington's personal apartments to the south and west, with his library occupying the southern corner, with windows to south-east and south-west, the sunniest place in the house. The library, where the preservation of books is an important requirement, has its long wall to the north-west, limiting the light level in order to preserve the books.

3.4 Science, Technology and Architecture in 19th Century Britain

The so-called 'Industrial Revolution' is generally dated to the period between 1750 and 1900, when new technologies were developed and brought to bear on almost all aspects of life, first in Europe and then throughout the developed world. It is generally accepted that these events had their beginnings in Britain (Derry and Williams 1960). In the field of building environment these new technologies were quickly applied to buildings in Britain and, as a direct consequence, a number of treatises were soon published offering detailed guidance on the design of systems for heating and ventilating buildings.[12] An important aspect of these developments is that, from the very beginning, architects of great eminence incorporated new environmental technologies in their buildings. To illustrate something of the work in Britain, the following illustrates buildings from the beginning, middle and end of the 19th century by three of the most important architects of the time, Sir John Soane (1753–1837), Sir Charles Barry (1795–1860) and Charles Rennie Mackintosh (1868–1928).

[12]For example Tredgold (1824, Richardson 1837, Bernan 1845, Reid 1844).

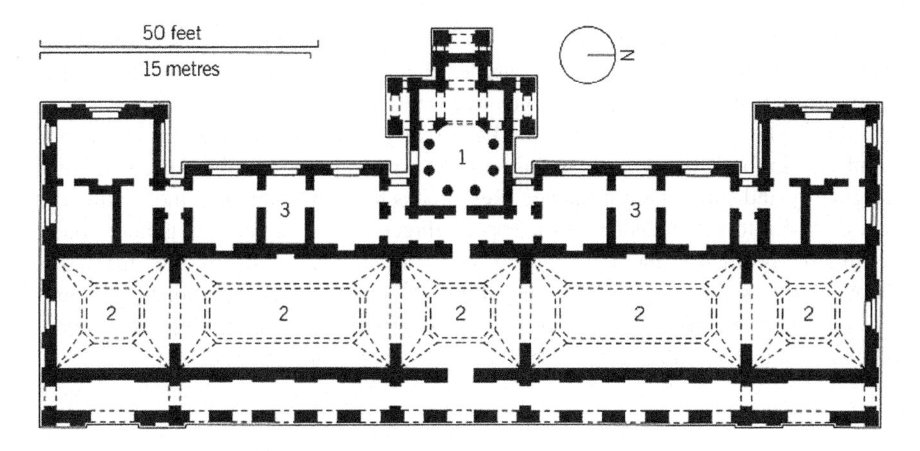

Fig. 4 Dulwich picture gallery, plan

3.5 *Building Four: Sir John Soane, Dulwich Picture Gallery, 1811–1813*

Dulwich lies some 10 km to the south of central London. At the beginning of the 19th century it was a rural village. Soane's Picture Gallery (Fig. 4) was built to house 360 paintings that had been bequeathed to Dulwich College. It was the first purpose-built art gallery open to the public in Britain and holds an important position in the subsequent history of the architecture of the art museum. In plan the building is a simple rectangle that is oriented with its long axis north-south. The pictures were displayed in five galleries *en fillade.* All lit by a 'monitor' rooflight that provided illumination of the pictures, without causing glare or unwanted reflections. The orientation ensures that only low-angle light from east and west enters the gallery spaces and south light is excluded, providing controlled light on the pictures and avoiding the heat and glare of the midday. This arrangement soon became a model for gallery design whose influence continued through the 20th century.[13] The building also included six small almshouses and, strangely, a mausoleum for the remains of the benefactor and two others.

The gallery was a key building in Soane's experiments with new means of space heating. The picture galleries were conceived to be heated by a steam-based system installed by the major engineers of the time, Matthew Boulton and James Watt.[14] A

[13]At the end of the 20th century Robert Venturi acknowledged the precedent of Dulwich upon his design for the Sainsbury Wing at the National Gallery in London. See Hawkes (1996).

[14]A comprehensive account of the Dulwich heating is Willmert (1993).

duct ran beneath the floor the entire length of the galleries containing the steam pipes to serve cylindrical heating batteries that stood in the centre of each space. The system was, thus, completely integrated into the architecture of this remarkable building. These spaces were a remarkable early synthesis of new environmental thinking, in both the visual and thermal, to produce a building of remarkable originality and subsequent influence. Researches have indicated that the heating did not extend into the mausoleum that opens directly from the central gallery space. Willmert speculates that this was to give thermal emphasis to the distinction between this solemn space and the galleries.[15] In its refined combination of precisely calculated lighting and complete integration of heating into the fabric, Dulwich is a key building in the history of environmental architecture.

3.6 *Building Five: Sir Charles Barry, the Reform Club, London, 1837–1840*

In the centre of the 19th century metropolis the matter of environmental design took on a new dimension. Unlike the semi-rural condition of Dulwich, the city centre was notoriously polluted. Although the city had endured a poor atmosphere since mediaeval times,[16] the problem became particularly serious following the enormous increase in population in the 19th century, but it was now possible to offer some resistance to its effects by applying the new environmental technologies.

Sir Charles Barry exhibited an interest in environmental systems in many of his projects, not least at the Palace of Westminster (Smith 1976). The Reform Club in Pall Mall (Fig. 5) is, perhaps, the most complete example of his ability to incorporate extensive systems for warming, ventilating and, now, artificial lighting into a building that is a consummate essay in the adaptation of historical precedent to new circumstances. The prototype for the Reform Club is the Italian renaissance palazzo, specifically Sangallo's Palazzo Farnese. In translating the precedent from 16th century Rome to the very different condition of 19th century London, Barry made a number of crucial changes. The most striking was to put a glazed roof over the *cortile* to create a space, the Saloon that is useable at all seasons of the year in this very different climate. Concealed in the fabric and details of the building is a comprehensive network of supply and extract ducts that delivered warmth and fresh air to all the major public apartments and conducted away the fumes and heat produced by the gasoliers that provided illumination.[17] The installation was designed by a Dublin-based engineer, John Oldham, in collaboration with the

[15]Ibid.

[16]This is discussed at length in Brimblecombe (1987).

[17]For an extensive description see Olley (1985). The building is also discussed in a wider context in Hawkes (2012).

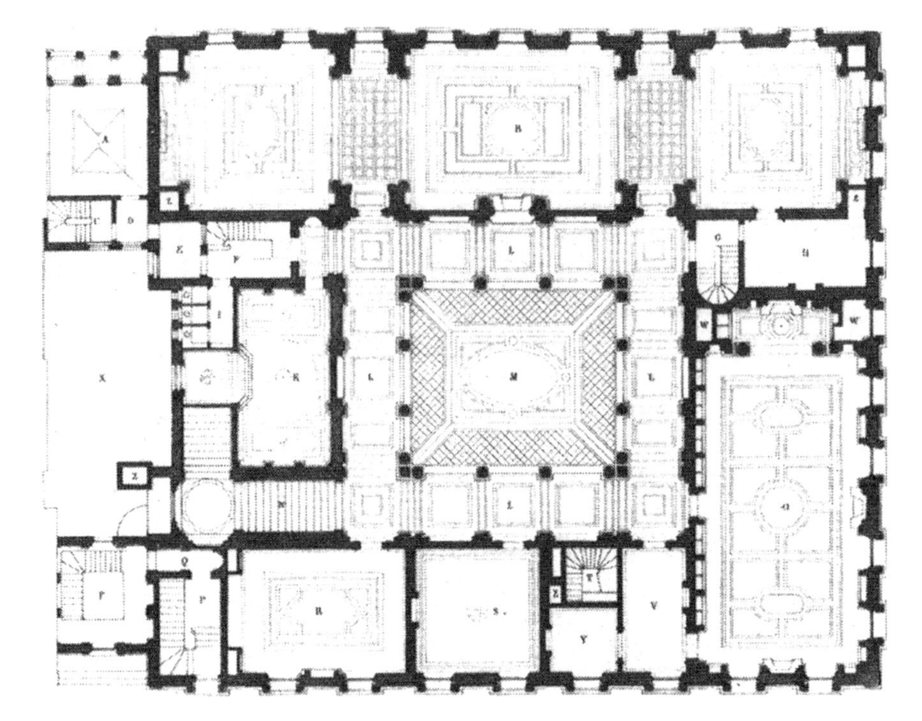

Fig. 5 Reform club, first floor plan

London firm of Manby and Price. It was powered by a five-horsepower steam engine, located under the pavement of Pall Mall. Incoming air was passed over a large heating coil and distributed throughout the building. The plan of the building beautifully responds to the orientation of the site in which the entrance is from the north on Pall Mall and the principal rooms are placed on the south, where they overlook Carlton House Gardens and enjoy the warmth of the sun.

3.7 Building Six: Charles Rennie Mackintosh, Glasgow School of Art, 1897–1909

At the end of the 19th century extensive mechanical systems for heating and ventilating were now commonplace in buildings of many types and the potential of electric lighting had quickly become appreciated. In that respect the Glasgow School of Art (Fig. 6) was relatively conventional. Designed by Mackintosh when he was a 29 year-old assistant in the Glasgow office of Honeyman and Keppie, the building merits its place in this discussion because of its remarkable synthesis of

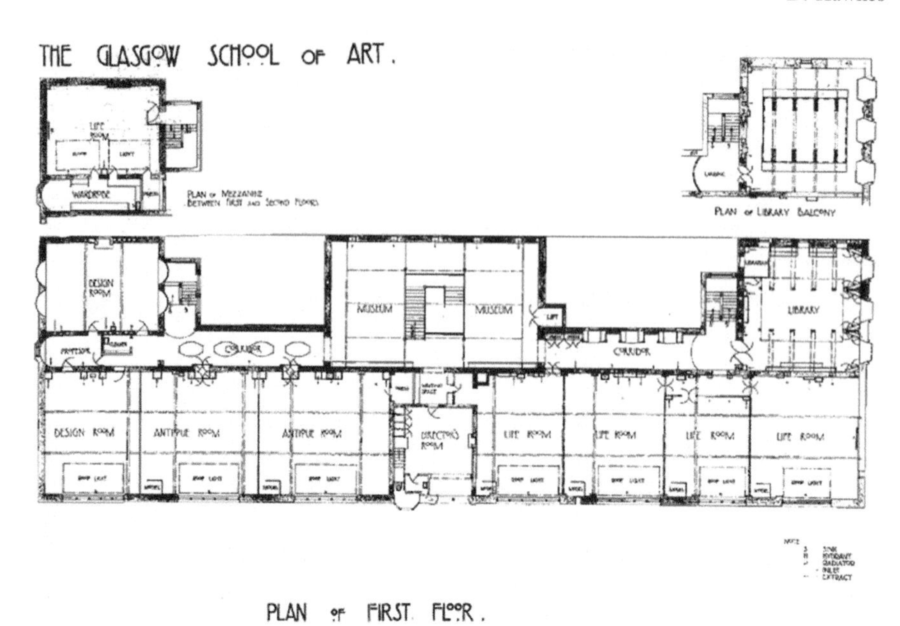

Fig. 6 Glasgow school of art, 1st floor plan

well-tested technologies and the greatest architectural invention to create an environment of remarkable quality and originality (Hawkes 2008).[18]

The dominant element of the building is the series of seven large painting studios that dominate the north-facing street elevation. Beyond these lies a spine wall that runs the entire length of the building, beyond which are placed a variety of spaces that enjoy the southern orientation. In addition to its structural function the spine wall is an essential element of the environmental system with a series of vertical ducts that supply warmed air throughout the building from the extensive plant rooms in the basement. Banham commented that,

> The provision of such a system of hot air ventilation and heating was a necessary concomitant of Mackintosh's use of huge north-facing windows in these rooms and a humane provision where the life class is concerned, for Glasgow is a chill city for nude models. (Banham 1984)

This strategy is uncannily similar to Robert Smythson's relationship between the fireplace-filled spine wall and glassy perimeter at Hardwick Hall, three centuries earlier, although here additional heat is provided from radiators beneath the windows.

The genius of Mackintosh's design lies in the manner in which he finely calibrates the environmental qualities of the individual spaces to their specific uses. From the large, brightly-lit, utilitarian volumes of the painting studios, with their

[18]See also Lawrence (2014) for a detailed recent study of the building.

compensating inputs of heat, to the dynamic light that enters the magical, dark-panelled library through three enormous west-facing windows, warmed, again, by a combination of warm air and radiators. Then there is the sun-filled loggia that hovers above the city on the upper floor of the south elevation, that is in effect a proto passive-solar space.

It is rarely observed that Mackintosh was a pioneer in exploring the potential of electric lighting in his buildings. The incandescent lamp was patented jointly by Swan and Edison in 1879 (O'Dea 1958). Within twenty years of this the Art School was lit after dark by a variety of light sources that ranged from utilitarian metal shades in the robust architecture of the studios and workshops to exquisite purpose-made fittings like the clusters of copper reflectors that light the Director's dark-panelled office and the astonishing chandeliers of cast metal and coloured glass that hang in the tall central space of the library.

4 Conclusion

We have travelled through three centuries of British architectural history, from Smythson to Mackintosh. In the 16th century science, in its 'modern' definition, was unknown. Parameters of the natural climate, temperature and barometric pressure were not yet measureable. Design therefore depended on the authority of precedent, of *tradition*. This is evident in Smythson's Hardwick Hall, where the starling originality of the building's first appearance reveals, on close inspection, continuity of tradition in the organization of its principal spaces. Just over half a century later, however, climate was measureable and we have the useful coincidence of Christopher Wren demonstrating both a 'Weather Clock' and a design for as building at meetings of the Royal Society. The connection between 'scientific' thought and architecture is thereby established and, I suggest, this is manifest in Wren's buildings, such as the Sheldonian Theatre. The 'Palladians' brought the architecture of 16th century Italy to 18th century England, in both theory and practice. They were, however, 'scientific' in making precise adaptations of the Italian model to the very different English climate, and in adopting the Newtonian practice of numerical tabulation to present their data.

In all these buildings the historic elements of architecture—form, fabric, plan and section—were the primary instruments of environmental management. In other words they remained rooted in historic practices. At the end of the 18th century, however, the events of the Industrial Revolution brought new means of generating and distributing warmth and air throughout buildings and, later, new methods of artificial lighting. Alongside these came new formulations for calculating the dimensions of pipes, ducts, heat exchangers and the like. This was a fundamental shift in the entire basis of architecture that changed the relation between buildings and climate. The foundations of future practice had been laid. On the other hand, the buildings of Soane, Barry and Mackintosh, whilst exploring the potential of the new tools and technologies to the full, combined these with an acute sense of the value of tradition. My conclusion is that this fact carries important implications for present day environmental design practice in architecture, where the benefits of new technologies of systems and materials and the power of our sophisticated, science-based design tools may be combined with the lessons of history, tradition and science.

References

Banham R (1984) The architecture of the well-tempered environment,, 2nd edn. The Architectural Press, London, 1984

Bernan W (1845) On the history and art of warming and ventilating buildings. G. Bell, London

Brimblecombe P (1987) The big smoke: a history of air pollution in London since mediaeval times. Methuen, London

Derry TK, Williams T (1960) A short history of technology. Oxford University Press, Oxford

Eliot TS (1920) Tradition and the Individual Talent', first published in The Egotist, 1919, later in The sacred wood: essays on poetry and criticism. Methuen, London

Giedion S (1948) Mechanization Takes Command: a contribution to anonymous history. Oxford University Press, Oxford

Girouard M (1983) Robert Smythson and the english country house. Yale University Press, New Haven & London

Hawkes D (1996) The environmental tradition. E. & F.N. Spon, London

Hawkes D (2008) The environmental imagination. Routledge, London & New York

Hawkes D (2012) Architecture and climate: an environmental history of british architecture, 1600–2000. Routledge, London

Hawkes D (2014) The origins of building science in the architecture of renaissance England. Wolkenkuckucksheim 19(34)

Hinchcliffe T (2004) Robert Morris: architecture and the scientific cast of mind in early eighteenth-century England. Arch Hist 47:117–142

Hopkinson RG (1964) Architectural physics: lighting. HMSO, London

Lawrence R (2014) The internal environment of the Glasgow School of Art by Charles Rennie Mackintosh. J Constr Hist 29(1)

Middleton WEK (1969) Invention of the meteorological instruments. Johns Hopkins University Press, Baltimore

Morris R (1734–1736) Lectures in architecture: consisting of rules founded upon harmonic and arithmetical proportions in building, London

O'Dea WT (1958) The social history of lighting. Routledge and Paul, London

Olley J (1985) The Reform Club. In: Cruickshank D (ed) Timeless architecture. The Architectural Press, London

Rapaport A (1969) House, form and culture. Prentice Hall, N.J.

Reid DB (1844) Illustrations of the theory and practice of ventilating. Longman, London

Richardson CJ (1837) A popular treatise on the warming and ventilating of buildings: showing the advantages of the heated water circulation. John Weale, London

Smith D (1976) The building services. In: Port MH (ed) The houses of parliament. Paul Mellon Center for British Art/Yale University Press, New Haven

Summerson J (1950) Sir Christopher Wren, PRS, 1632–1723. Notes and Records of the Royal Society, vol 15

Tredgold T (1824) On the principles and practice of warming and ventilating buildings. Joseph Taylor, London

Willmert T (1993) Heating methods and their impact on soane's work: lincoln's inn fields and dulwich picture gallery. J Soc Arch Hist LII:26–58

Wren C (1750) Junior, Parentalia: Memoirs of the family of the Wren, London

Not Just a Summer Temple: The Development of Conservation and Indoor Climate in Nationalmuseum, Sweden

Mattias Legnér

Abstract This contribution examines the building and management of Nationalmuseum in Stockholm, and explores how conservation and indoor climate was shaped by technological development and how views on the running of a museum building shifted.

> It is important that Nationalmuseum will become not just a summer temple, which strangers admire during their brief visits, but that it is a home to the fine arts, where we can be comfortable all year around, where we can stay and study even in the long and cold winters of Scandinavia (Sander 1866, 38).

1 Introduction

This essay examines the building and management of Nationalmuseum in Stockholm. Today the building has just recently been reopened after years of renovation and fitting of a new system that will control the indoor climate. This means deep interventions in a nineteenth-century building that was not designed to be airtight or to be heated all year around. The renovation gives a reason to ponder on how the building originally was designed and constructed, but also how it was managed over time. The climate of the house has been an issue ever since the building was constructed in the early years of the 1860s (Fig. 1). It was fitted with a central heating system already then, but the building proved difficult to heat in winter and to ventilate in summer. There were continuous problems with dehydration of organic materials in the art collections in winter-time, and with too much sunlight exposing fragile art in the warmer season. Curiously, the introduction of artificial humidification first around

M. Legnér (✉)
Department of Art History/Conservation, Uppsala University,
Campus Gotland, Visby, Sweden
e-mail: mattias.legner@konstvet.uu.se

C. Manfredi (ed.), *Addressing the Climate in Modern Age's Construction History*,
https://doi.org/10.1007/978-3-030-04465-7_7

1930 and then again in the 1950s did not solve the problem of dehydration. On the contrary climate problems became ever more complex around the mid-20th century because of the introduction of motor traffic with its exhausts, and increasing demands on a stable indoor climate in art museums. [1]

How did museums balance the needs of their collections, against the needs of staff and visitors? What considerations where made when choosing heating and ventilation for a museum at this time? In order to illuminate these questions, archival sources from Nationalmuseum, Riksarkivet (National State Archives) and the engineering and architectural company SWECO have been used. Överintendentsämbetet (Board of Public Works and Buildings, abbreviated ÖIÄ) was the custodian of government buildings, followed by Kungliga Byggnadsstyrelsen (Board of Building and Planning, abbreviated KBS) after an organizational shift in 1918. The museum was thus responsible for the management of its collections but not of its building. Until 1939 there was also a second museum housed in the bottom floor: Statens Historiska Museum, the National Historical Museum.

If the museum had a complaint on the performance of the building or the heating system, it would have to notify ÖIÄ (or KBS after 1918), which then would decide how to act. Judging by archive sources, it becomes evident that ÖIÄ had small means to make more demanding interventions in existing buildings, and often complaints seem to have been more or less ignored because of lack of resources. By studying the correspondence it is possible to gain a better understanding of how museum management perceived indoor climate and how ÖIÄ responded.

The purpose of the essay is to explore how the construction and management of the indoor climate was shaped by technological development and how views on the running of a museum building shifted. Nationalmuseum was fitted with a hot water central heating system. In the early 1860s this was something hardly heard of in Sweden at this time. In general, the central heating systems used at that time were caloriphers, furnaces that heated the air that was then circulated through the building.

There were firms in Stockholm installing piping, but none of them was considered competent enough to do the installations in Nationalmuseum. Most entrepreneurs in Stockholm worked with gas piping, not with water or sewer piping.[2] In the early 1860s it was still not evident that a public building should be equipped with this kind of heating, despite the relatively long and cold winters in Stockholm. Public buildings in general were heated with local fireplaces, most often tile stoves produced in the city.

Today it is well known that control of indoor climate is key to the management of collections. Too much heat makes the air dry, which may cause damage to fragile objects such as paintings on panels or wooden furniture with veneer. Too little heat makes the air very humid, which promotes mold, vermin, corrosion and rot. What is considered “too little” or “too much”, however, has changed since the nineteenth

[1]Legnér & Geiger (2015)

[2]Stålbom (2010). In 1861 Stockholm opened its first waterworks with 30 km of piping.

century.[3] The essay explores why central heating was installed in the museum, what the expectations on its functioning were, and how building and museum management (they were—and are—separate from each other) continuously commented on its performance in the decades following the opening of the museum, up until the 1970s when air pollution had become a serious problem demanding a technical solution.

2 The Decision to Build a Museum

Museums were not alien to Swedish culture of the mid-nineteenth century, but they were few and almost none had a building that had been designed to house a museum. The best museums were the ones belonging to the universities in Lund and Uppsala, where collections had been accumulated over a long time. Swedish architects and artists knew especially German, French and Italian museums. A common problem in Western and Northern Europe was how to use sunlight most efficiently in museums and libraries. European museums of the nineteenth century were generally dark, gloomy places which had very limited hours during winter time since they could not be heated efficiently and were completely dependent on natural lighting. Kunsthalle in Karlsruhe, for example, was open for two hours per day in wintertime, while museums in Berlin where open for four hours in the middle of the day.[4]

There had been voices calling for a new national museum building in Stockholm since the early nineteenth century, when parts of the royal collections were transferred from having been royal property to becoming state property. The collections were kept in a number of places inside and outside of Stockholm that were not easily accessible to the public, such as Gripsholm Castle, Riddarholmen Church and the Drottningholm and Ulriksdal Palaces. At least a couple of these places were clearly not suitable to house collections because of humidity and lack of heating.[5]

At the time (1845) when a decision was made to build Nationalmuseum, it was an astonishingly expensive project for state government. Around mid-century Sweden was still an agrarian economy of which the vast majority of the population lived outside towns and made their income on agriculture, forestry, fishing and some handicrafts. Using public funding to erect a monumental building in the capital that would celebrate Swedish arts and history was highly controversial. In the intellectual and political debates preceding the decision to build the museum the first seeds of modern Swedish cultural policies were planted. Per Widén has argued that there was in fact some support for an art museum in Stockholm in the 1830s, despite that the parliament had voted against a national museum as late as 1828.

[3]Legnér (2015).
[4]Sheehan (2000).
[5]Legnér (2011).

Critics were generally not hostile to the idea of a museum, but they were wary of the costs associated with its building. After the parliament's dismissal the king Karl XIV Johan went ahead with his own plans for an art museum next to his summer palace Rosendal on Djurgården in the outskirts of Stockholm. Djurgården was a large, forested area disposed by the king.[6] Today it is a recreational area for Stockholmers and the home of a number of government institutions, among them a number of national museums, to mention a couple of its uses.

The issue of building a national museum divided the estates of the parliament: peasants and burghers opposed the project fervently, whereas nobility and clergy voted in favor of it. There was no doubt that this was a project that found most of its support in the king and the nobility. Some opponents claimed that the sum proposed was far from sufficient to design, construct and furnish such a building.

Nonetheless, since Riksdagen (the Swedish parliament) could not come to a decision it was instead decided by an extended committee of the museum project that 500,000 Riksdaler Riksmynt—a huge sum for anyone at the time—should be allocated to erect Nationalmuseum.[7] It would become a huge investment for the Swedish state. Half a million Riksdaler Riksmynt would not suffice by far. In the end the construction would cost 4.5 times this sum.[8] It was the most expensive building project since the erection of a new Royal Palace in the eighteenth century (the old one had been ravaged by fire in 1697).

3 Building the Museum

An initial object of concern was the location of the museum. There was lobbying for different locations. One party wished to see the museum erected very close to the Royal Palace and other government buildings on Helgeandsholmen. A military captain called Baltzar Cronstrand wrote several articles in the daily press, calling for a central location. The city of Stockholm wished to see the building erected on Kyrkholmen, which was a slum area not very developed but relatively close to the government quarters and the gentrifying northern part of the city. The location was furthermore facing the Royal Palace on the opposite side of the Norrström water. It is not hard to see why the city wished to see this area developed with a monumental public building. The costs for building there would be significantly lower than building on Helgeandsholmen, which was a densely developed area and poorly accessible site. With such a location it would have been very hard to keep the allocated budget. It seems as if the location on Kyrkholmen was chosen mainly because it was a poorly developed area close to the city center.

[6]Widén (2009).

[7]Malmborg (1941).

[8]Sander (1866).

Construction was begun surprisingly quickly, following a proposal made by the young and inexperienced architect Fredrik Wilhelm Scholander. It has been discussed why ÖIÄ went ahead with a proposal which evidently was incomplete. The board had not taken into account the size of the collections to be kept in the building. Corruption might have been one reason.[9] This rash decision soon proved to be a mistake and construction of the foundation was interrupted after a year.

The design was heavily influenced by requirements not to exceed budget. When the location of Kyrkholmen had been decided by the King the superintendent of ÖIÄ and military officer Michael Gustaf Anckarsvärd, an architect who was the overseer of government buildings, proposed the dimensions of the building. It would have three floors and two courtyards illuminating the galleries. At this point the floor area allocated to different collections was decided after negotiation between the involved institutions. The art collections would occupy less than a third of the space, the royal library about half, the historical collections one seventh and the royal armory less than a tenth, which would make it very cramped. As the committee of the building project approved Anckarsvärd's promemoria, these basic dimensions (given in absolute numbers) became another object of concern.

The challenge of erecting a monumental building which could house six collections managed by three institutions, and keeping within the allocated budget, was immense. Never the less was this a highly prestigious mission for the superintendent's office which was in dire need of improving its reputation within public government. Military engineers who were closely connected with the royal court challenged from the outset the superintendent. Several military officers who had the King's ear, among them Baltzar Cronstrand and also Johan af Kleen, complained vigorously over Anckarsvärd's choice to let a young, just graduated architect draw the building. af Kleen had his reasons to complain, since he would be appointed as the new architect after the King had dismissed Scholander. The scholar Bo Grandien has argued that the disapproval of the proposal was directed not towards Scholander's person, but against his party—namely the Board of Public Works and Buildings (ÖIÄ).[10]

The king, influenced by critics in his entourage, now acted to marginalize the role of ÖIÄ. Suddenly, Scholander was asked to make a trip to Berlin in order to study museums. After a few days Anckarsvärd received new instructions from the king. Despite that Scholander remade his drawings to suit the king's wishes, he was soon sidestepped to the benefit of Johan af Kleen, who was the King's favorite Kleen would now be the one in charge during the trip to Berlin. Scholander's role was to accompany him in order to gather information for af Kleen. The object of the trip was now to consult the architect August Friedrich Stüler, who had been a pupil of the museum architect Leo van Klenze and who had recently designed the picture gallery Neues Museum. Shortly after the trip the trajectory of the project changed completely. Both af Kleen and Scholander were dismissed, and Stüler's position

[9]Grandien (1976).

[10]Grandien, "Det Scholanderska fiaskot".

quickly shifted from having been an outside consultant to becoming the museum architect. Not only did the group of actors change but also the objects of concern. True, the foundation of the museum had already been laid, but Stüler was not instructed to keep within the specified budget. Suddenly, with an architect of international class present, there was no longer an obvious need to follow directives given by parliament. Riksdagen, which in 1846 had taken the initiative to erect the building, had by the summer of 1847 become completely marginalized by the king's party.[11] With Stüler the project could cleanse itself of an infected conflict between a fraction of military architects and ÖIÄ. Furthermore it received a competent and proven architect who was not easily criticized. By engaging a famous foreign architect the king hoped that the debate on the design of the museum could be put to an end and his wish be carried out.

Much like other museums of the nineteenth century Nationalmuseum was built at a time when museums were still viewed as mere galleries. Collections would be put on view, but in Sweden there was not the notion that museums also needed storage space, ample working space for staff and also workshops. The monumentality of the building—its potential of communicating national pride to the viewer—was considered to be of immense importance. August Stüler followed a tradition of German architecture when he argued that the building in itself should be a work of art with its character borrowed from "the dominant period of art history, such as "the fairest times of antiquity or the purest and most original period of Italian Renaissance."[12] This would be a *national* museum, not just a museum of art. The intention was that all of the royal collections would be kept and displayed in the building. The plan was to bring a total of six collections together under one roof: The Royal Museum, The Royal Swedish Academy of Letters, History and Antiquities, The Royal Coin Cabinet, The Royal Clothing Chamber, the Royal book collection and the Royal Armory.[13] In the end, the Royal book collection was never moved here but was installed in a new building of its own in 1874. Instead Nationalmuseum became a refuge for the royal art collections.

There had been complaints about the unsuitable environments of the Royal Palace and the Riddarholmen Church, which was the official burial site of the royal families.[14] In the palace paintings had been subjected to dust and the flames and smoke from candles and torches, while moist air was pointed out as an environmental problem in the church. In 1866 they were in need of a skilled conservator. The Riddarholmen Church was the home of many war trophies from the 17th century but was a very damp and cold environment, clearly not suitable for the display of armor, flags, drums and arms.

---

[11]von Malmborg, "Nationalmusei byggnad", 57.

[12]Laine (1976). Author's translation.

[13]von Dardel (1866).

[14]von Dardel (1866).

Fig. 1 The opening of Nationalmuseum in 1866 was big news in Sweden. *Archival source* Konstbiblioteket, Stockholm

4 The Perkins System and Indoor Climate in Nationalmuseum in the Nineteenth Century

We do not know much about the actual reasons for installing central heating in the building, but one reason seems to have been to protect national treasures. In the budget for the building of 1849 there was no cost included for a heating system, but ten years later funding for a central heating system had finally been approved by the parliament following a debate on what sort of system—hot air or hot water—should be selected.

In August 1859 the sum of 50,000 Riksdaler Riksmynt had been earmarked for heating devices.[15] It is not entirely clear why Perkins ovens were selected in the end. What is clear is that the heating system was not viewed as part of the building construction, but as part of the furnishings. Funding for the interiors such as windows, floors, doors, and piping was not allocated until 1860, at which time the costs for building the museum seemed to have gone completely out of hand. One feature affecting the indoor climate that was not chosen because of budget

[15]von Dardel (1866).

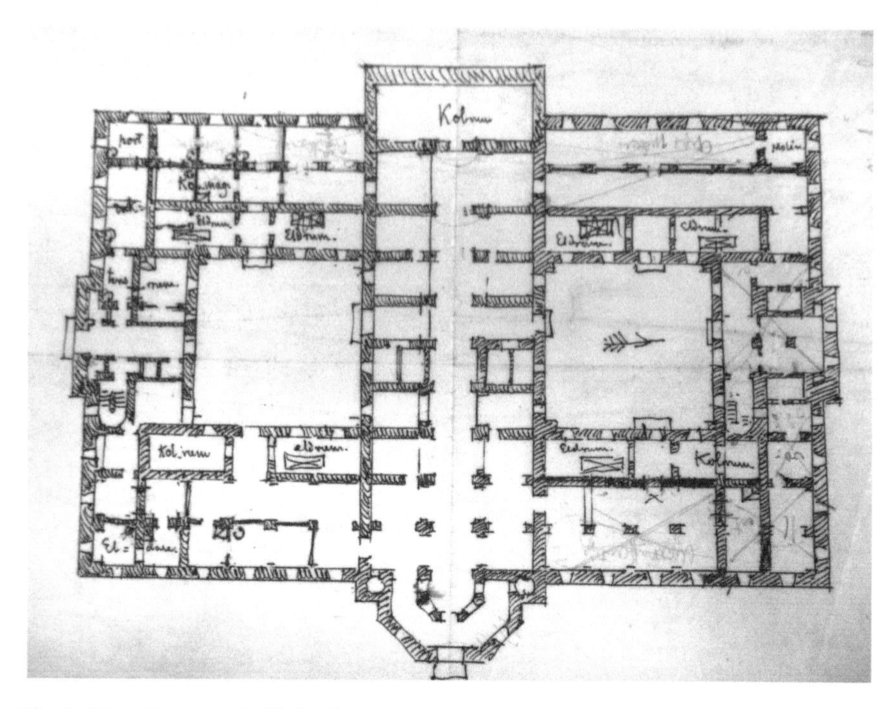

Fig. 2 Plan of basement in Nationalmuseum, 1860s. A large part of the basement was taken up by furnaces and storage for coal and firewood. There were six furnaces ("eldrum"), four rooms for coal ("kolrum") and living quarters for the stoker and his family. *Archival source* Riksarkivet, Överintendentsämbetet, vol. DIIba: 3, plan S132

limitations was secondary glazing of windows. According to a calculation from 1859 this would have cost 30,000 Riksdaler Riksmynt, and was considered too expensive at this stage.[16]

The costs were scrutinized by the state committee, which wrote a proposal for a budget to be decided by the parliament. The proposal was debated quite lively among the estates in September 1860. A number of members of the peasants' and burghers' estates were upset about the uncontrolled rise in costs. Now the building committee wanted the parliament to allocate additional funding for furnishing the interiors of the building. As part of this debate the burgers' estate discussed the choice of heating system, since this choice could have some effect on the budget. One member questioned the state committee and ÖIÄ's recommendation of a hot water system. He argued that a caloripher (hot air system) would be more affordable and also produce a healthier indoor climate. A caloripher would also dry out the moisture of the building quicker. Another member, Lars Hierta, partly conquered but asked if the sum of 60,000 Riksdaler Riksmynt was not set too high. Could not tile

[16]Riksarkivet (RA), Överintendentsämbetet (ÖIÄ), FIab: 55, "P.M." augusti 1859 (antagligen avskrift).

stoves produce heat just as well and be much cheaper to install? However, as a conclusion of the debate the estate voted for approving the proposal of the state committee, which implies that there was not much critique against the budget.[17]

In order to understand the questions asked by the members of the burghers' estate, it is necessary to know that issues of heating had not yet acquired the exclusively technical character that they were given towards the end of the century. They could obviously become issues of political debate and were not discussed just by engineers or architects. Installing central heating in public buildings was not yet a common measure, and could be criticized for being a waste of public funds.[18] Local fireplaces still dominated the heating of both private and public buildings.

In 1861 the construction work had progressed to the degree that the building committee was dissolved and replaced by a committee responsible for furnishing the interiors, including the fitting of a central heating system and tile stoves. This committee would mostly work with the issue of how to divide space in the building between the different collections and institutions involved. The committee consisted of representatives of these institutions: the superintendent of ÖIÄ Gustaf Söderberg, the curator J. Chr. Boklund, the custodian of antiquities (*riksantikvarie*) Bengt Emil Hildebrand and the librarian Gustaf Erik Klemming. The committee was led by the courtier and poet Gunnar Wennerberg. Since the committee could not reach consensus on how to divide the spaces, Wennerberg finally made a proposal of his own. His proposal meant that sculptures and trophies would be placed on the middle floor, antiquities on the ground floor and paintings on the top floor. As a consequence the book collection could not be housed in the museum but would instead need a building of its own.[19]

In 1862 the German firm of Johannes Haag in Augsburg at installed the Perkins system at a cost of 61,000 Riksdaler Riksmynt. The Perkins ovens meant that hot water would gravitate through a system of one-inch thick pipes.[20] The pipes passed through a brick oven located in the basement. The network of pipes was fitted with an expansion loop. When the water expanded through heating it filled the loop, making it possible to heat the water beyond its boiling point. Water reached temperatures of 150–200 °C. The advantage of this system was that it allowed much thinner pipes that could more easily be concealed than what low-pressure hot water heating needed at this time. The Perkins oven could achieve the same heating capacity as larger pipes of earlier hot water systems.[21] A considerable drawback, however, was that the water was to be heated far beyond the boiling point in order

[17]*Borgarståndets riksdagsprotokoll* 1859–60, vol. 6, 363–375 (22 September, 1860); the statement of the state committee in *Bihang till riksdagsprotokoll* 1859–60, vol. 11, no. 180.

[18]This was however about to change very soon. When the royal library, *Kungliga Biblioteket,* was planned a few years later central heating was seen as necessary at least in the public areas of the library.

[19]Bjurström (1992). See also Willers (1977). The royal library opened in 1874.

[20]For an exhaustive description of the application and dissemination of Perkins systems see Manfredi (2013).

[21]Willmert (1993).

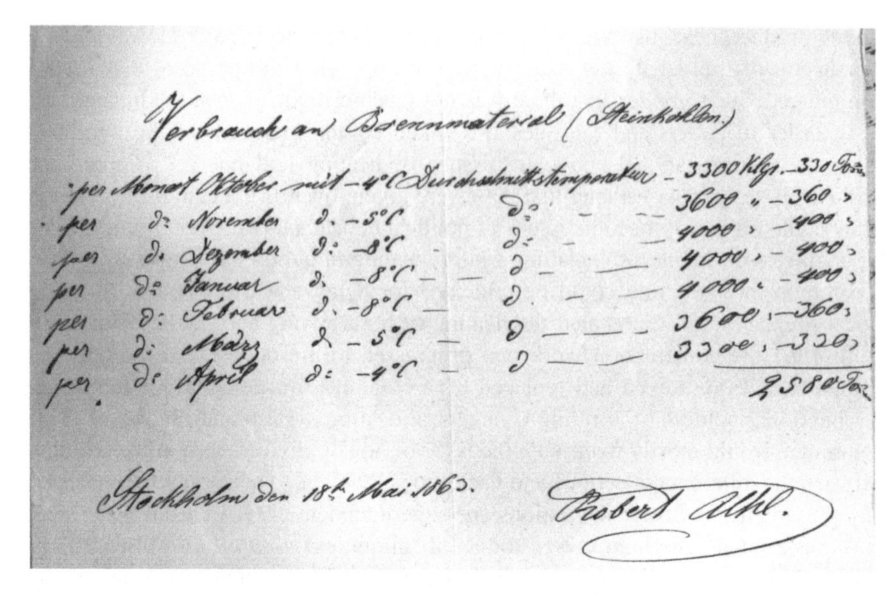

Verbrauch an Brennmaterial (Steinkohlen)

per Monat Oktober mit -4°C Durchschnittstemperatur – 3300 Klgr. – 330 Ton.
per d: November d. -5°C d: – 3600 " – 360 "
per d. Dezember d: -8°C d: – 4000 " – 400 "
per d: Januar d. -8°C d – 4000 " – 400 "
per d: Februar d. -8°C d. – 4000 " – 400 "
per d: März d -5°C d – 3600 " – 360 "
per d: April d: -4°C d – 3300 " – 320 "

2580 Ton.

Stockholm den 18t Mai 1863.
Robert Ahl.

Fig. 3 One of the galleries in the beginning of the 20th century. There were heating pipes in the floor, covered by a protective grate (*Sveriges Nationalmuseum i bilder*, 1906)

to increase pressure, making it possible to raise pressure to dangerous levels where an explosion could occur.

Perkins ovens, first developed by the American born engineer Angier March Perkins mainly working in England, was installed in a number of public buildings in England in the 1830s. In 1835 the British Museum installed two of them, one for the reading room and another for the bird and print rooms.[22] In 1833 a Perkins system was installed in the Court Room of Bank of England, and the Law Courts at Westminster is also supposed to have had a Perkins oven.[23] Perkins systems have also been installed in many churches in England.[24]

In order to heat all the galleries in Nationalmuseum a total of six ovens were considered necessary by the entrepreneur. During operation (daytime only) they would be constantly fed and supervised by two stokers. In addition to that, a caretaker living in the building was contracted to operate the ovens and see to that they were not overheated. The caretaker was a mason who had been involved in putting the pipes of the Perkins ovens into walls and joists.[25] Most of the basement was occupied with functions relating to the heating system: the ovens, fuel storage, and living quarters for the caretaker (Fig. 2).

---

[22]Ierley (1999).

[23]Hawkes (2012).

[24]See the CIBSE Heritage Group Website: http://www.hevac-heritage.org/victorian_engineers/perkins/perkins.htm#perkins2.

[25]RA, ÖIÄ, vol. FIab: 56, dnr 288, 5 April, 1866.

The museum was heated in daytime from September and well into May. There were repeated accidents with the oven: if the water was heated too much, pipes would burst and need to be replaced, and before spare parts had been obtained from Germany and put into place the ovens would not work. In the first years there was a supply of spare parts available in the museum, but this supply was not replenished and since there were many repairs it eventually became empty.

There were no radiators, but instead hot water was lead through seven kilometers of pipes embedded in the wooden floors. Heat from theses pipes rose through grates in the floor (Fig. 3). The hot water pipes were causing damages to the parquet, and the inside of the ducts had to be covered with plate to protect the floor.[26]

A representative of the Johannes Haag Company, an engineer named Robert Uhl, conducted the first tests of the system. The system was started in October 1862 and was used until the end of April 1863. This was considered to be a normal annual use of the system. In protocols from these tests there is information about fuel consumption and *durchsnittstemperatur,* i.e. average temperature. Exactly what is meant by average temperature here is not explained since there is no information on exactly where temperature was measured (Fig. 4). It is possible that average temperature was considered an overall average based on measurements on all floors. In the middle of winter 400 barrels of coal were used every month, and an average indoor temperature of 8 °C was reached.[27]

The system was vulnerable. During tests there were several explosions caused by too high pressure in the pipes.[28] The weakest parts were the joints that could easily break if the pressure became very high. Water with a temperature of upwards 200 ° C would then spray directly into a gallery or leak into the masonry walls. When there was leakage the system could overheat in just a few minutes, breaking the pipes inside the ovens. Later this risk of overheating would become part of the critique aimed at the Perkins system of Nationalmuseum.

The "heating apparatus" was evaluated and improved several times after the installation had been made.[29] A first independent evaluation was conducted in 1863 by consultants employed by ÖIÄ.[30] The average temperature was said to be 15° above outdoor temperature at the beginning of the test. Temperature rose quickly during the test. At 25 °C it was decided that the test should be cancelled because of the risk of damaging the wooden floor that had recently been put in. The test showed the powerful heat that the system could give off. It also shows that those involved were conscious of the influence of temperature on the humidity of air.

Consequently, this system was not perceived as optimal even at the time of its introduction. The fuel consumption was about 25% higher (3.2 barrels of coal

[26]Nationalmuseum (NM), protocols in museum matters, vol. A2: 4, protocol 8 April, 1869.

[27]RA, ÖIÄ, FIab: 55, R. Uhl's calculation 5 December, 1863. 400 barrels should have been equivalent to 60,000–68,000 l, depending on which kind of barrel was used.

[28]RA, ÖIÄ, FIab: 55, utan datering.

[29]RA, ÖIÄ, FIab: 55, R. Uhl 17 June, 1863, and K. Styffe 18 February, 1864.

[30]RA, ÖIÄ, FIab: 55, nr 1863-06-16.

Fig. 4 Table of fuel consumption of the Perkin's system during the trials in 1862–63, showing the huge volume of coal used for heating the building in winter-time. *Archival source* Riksarkivet, Överintendentsämbetet, vol. FIab: 55

Fig. 5 The vestibule of Nationalmuseum. Image printed in *Ny illustrerad tidning* 1866, showing that visitors entered the exhibits fully dressed, only leaving canes and umbrellas at the entrance

instead of the specified 2.6) than predicted by the manufacturer, supposedly due to the large single pane windows through which a lot of heat was lost.[31] In the first year of service the annual budget for fuel roughly equaled the salary for the three staff members maintaining the heating system and the costs for its repairs.[32] About sixty years later, in 1923–24, the annual expenses for heating the building were 17,009 Kronor. A mix of birchwood, softwood, coal and cokes were used for fuel, with cokes as a dominating part.[33]

In 1865 it was decided to move 888 paintings from the Royal Palace, 67 from Drottningholm Palace, and 22 from Gripsholm Castle, which were royal castles with large repositories of paintings.[34] The overarching purpose of this move was to give the collections a representative building, designed as a work of art itself, and make them available to the public for the first time. The move also had a symbolic meaning with the public government taking over the stewardship from the monarch. The collections that were moved to Nationalmuseum consisted only to a small

[31]RA, ÖIÄ, vol. FIab: 5, report by Styffe, Bolinder and Edlund, 8 February, 1864.

[32]von Dardel (1866).

[33]NM, protocols in museum matters, vol. A2: 59, form for fuel consumption.

[34]Sander (1876).

extent of paintings: rune stones, sarcophagi, liturgical objects, archaeological findings of gold and silver, royal dresses, about 100 sculptures, arms and armor, porcelain, sketches and models, war trophies, 12,885 drawings and 60,000 engravings were also to be housed in the new premises.[35] In 1884 trophies, arms and armors were moved to the Royal Palace in order to give room for the decorative arts.

The building housed Nationalmuseum, which managed the art collections, and the Royal Swedish Academy of Letters, History and Antiquities (subsequently called the Historical Museum) that kept historical and archaeological collections in the ground story and basement of the building. The art collections were kept on the upper two stories where lighting was better. Before 1930 the exhibition spaces had no electrical illumination.

Although source material commenting the indoor climate is scarce, there is evidence showing that objects brought into the museum quickly were damaged. The reason was that they had been stored in a humid and cool environment before they brought to a centrally heated building in which relative humidity should have been comparatively low. The Järstad triptych, which is a medieval wooden triptych, was placed in the custodian Hildebrand's office on the bottom floor (close to the ovens kept in the basement) of the museum, after which "the wood dried out, and the paint, applied to a gypsum primer, fell off in pieces and was badly damaged."[36]

A temperature of about 14 °C seems to have been the norm for the galleries in wintertime, but it was probably lower during the coldest part of the year. According to the first trials, the system should manage to raise the temperature 16 °C above outdoor temperature. The top floor with its skylights and insufficient heating sources should have been considerably colder, not to mention the attic where there was an outdoor climate. Early testing of the Perkins oven also showed that temperature varied between different parts of the building, depending on wind speed and direction.[37] Keeping a steady temperature was not possible in any part of the building, neither in the 1860s, nor fifty years later. In 1916, for instance, the Director of Antiquities complained to the Superintendent about the chill in the offices heated by tile stoves only. Around this time the heating system was supplemented with a few electrical radiators, and a steam boiler was installed in the basement.[38] Steam heat was used mainly to keep ice and snow off the skylights and not for thermal comfort. However, this circuit did not emit enough heat to keep the snow from building up. Skylights had to be shovelled and cleaned continuously in winter.

The Perkins system heated the galleries but did not serve the office spaces. In fact, the only reason why we know anything at all about temperatures in the building is that the staff complained about the level of thermal comfort. Tiled stoves

[35]von Dardel (1866).

[36]Quote from Nils Månsson Mandelgren in Stavenow (1972).

[37]RA, ÖIÄ, vol FIab: 55, report by Edlund and Folks, 29 December, 1864.

[38]RA, ÖIÄ, vol FIab: 60, folder on steam heating, 9 September, 1915.

but also gas stoves were installed in the rooms of the curators and their staff, but in wintertime it could still get uncomfortably cold. The Director of Antiquities Hans Hildebrand complained about the cold in his office on the ground floor, which had a temperature of 10–12 °C in wintertime.[39] In 1905 there were three studios there, but because of the chill the antiquarians could not work there.[40]

Storage spaces were not planned to fit into the building, but since the collections were too large for the building the attic and parts of the basements—premises were you would find the coolest, the most humid and the least stable climate—were taken into use for these purposes. In 1863 it had become apparent that some kind of storage space would be necessary in a near future, and for that reason a floor had been fitted in the attic.[41] Some objects were moved down into the basement, which was a very damp place (the floors rotted away) due to the proximity of the lake Strömmen, located just next to the building and almost at the same level.[42] A complaint from the fire department in 1915 about the fire hazard in the attic proves that it was still used for storing paintings and materials for restoration work. Also after the complaint in 1915 was the attic used for storage purposes.

The by far worst working conditions to be found in the building was the restorer's workshop (furnished in 1873) in the attic, where the temperature reportedly was between 0° and 6° in winter close to the stove. Already at this time it was seen as a fire hazard to have a stove burning continuously in winter in the attic where a fire could spread rapidly, and where few staff members moved about, but none the less the attic continued to be used as a restorer's workshop with an outdoor climate. In the 1890s the floor was filled with brick in order to make the attic more fire proof, a measure that should have made the attic even chillier.

Evidently there was a conflict between the wish to preserve the collections, and the demand for comfortable climate that was, if not comfortable, at least bearable in winter. Offices were heated with tile stoves that during cold winter days were fired constantly. Apparently this was not enough if one wanted to establish a climate appropriate for the working conditions.

In short, Nationalmuseum before the 1920s was a rather chilly place compared to later comfort standards. Visitors entered the exhibits fully dressed, only leaving canes or umbrellas to the porter (Fig. 5). It was heated only during working hours, allowing the building to cool off during the night. As a consequence, large amounts of water would condensate on the inside of skylights and windows in night-time. This water had to be led away to large tanks in order not to flood the galleries on the third story. One single winter night could produce as much as 1,000 litres of condensed water inside the building.[43]

---

[39]RA, ÖIÄ, vol FIab: 58, letter from Hildebrand, 8 December, 1902.

[40]RA, ÖIÄ, vol FIab: 58, staff letter, 25 February, 1905.

[41]RA, ÖIÄ, vol FIab: 55, inspection report, 16 June, 1863.

[42]RA, ÖIÄ. On fire hazards: vol. FIab: 59, staff letter 16 August, 1915. On rot in the basement: vol. FIab: 57, staff letter 26 June, 1889.

[43]RA, ÖIÄ, vol. FIab: 55, ref. 48, 16 February, 1864.

The central heating system was primarily intended for the galleries, and this is why additional heat sources such as tile, gas and later also electrical stoves had to be used in other spaces. There was little possibility of controlling the indoor climate in the building, even though the Historical Museum obviously attempted to add moisture to the air by putting open water containers in the exhibitions, and there was also the risk of pipes bursting or a fire breaking out.[44] In 1919 the Director of Antiquities Otto Janse complained to the Superintendent about excessive heating on the first story, where the collection of medieval polychrome wood was kept. He demanded that the pipes be insulated since attempts to decrease the water temperature in the pipes had not worked. The pipes in the floor of the gallery were insulated with asbestos paper, but apparently this did not decrease air temperature much.[45] Instead the Historical Museum tried to humidify the air by putting out containers of water. When temperature was raised throughout the building after 1930, the problem of dry air would become more serious.

5 Demands for Electrification and a New Heating and Ventilation System

An incident in 1923 made some of the problems with the existing heating device painfully apparent. The walls of the French gallery had recently been repainted, but soon it became clear that the hot water pipes dirtied the walls and the ceiling, making the newly renovated gallery look old and grimy again. Curator Axel Gauffin made an attempt to reduce the blackening effect by installing an oak bench on the walls, just above the conduits in the floor.[46] This had little, if any, effect. Dirt just stuck to the wall harder and higher up, and after an accident where a visitor had stumbled on the bench and damaged a painting, the bench was removed. This, and surely also other circumstances, made Gauffin officially condemn the Perkins oven as being hopelessly outdated and unsuitable for a modern museum building:

> … the outdated heating device of the building, as unsound for the visitor and the staff as for the objects of art, and whose replacement with a new system will prove necessary in short, also for the reason that with the existing system heating the museum in the evenings is impossible. Because of this, an obstacle for the progression of the museum into an educational institution for the people has been laid.[47]

[44]RA, Kungl. Byggnadsstyrelsen (KBS), Intendentsbyrån, vol. F1A: 116, letter from O. Janse, 18 February, 1919.

[45]RA, KBS, vol. F1A: 116, letter from O. Janse, 18 February, and letter from H. Theorell, 11 March, 1919.

[46]NM, protocols in museum matters, vol. A2: 58, 3 May, 1923.

[47]*Meddelanden från Nationalmuseum nr 48,* Nationalmuseum, Stockholm 1923, 4. Author's translation.

The argument, thus, was that the existing device was a risk both for collections and humans. If Nationalmuseum was to become an institution attractive and available to the public, it should also offer a comfortable environment.

The campaign for a new heating system that Gauffin had initiated in 1923 was continued in the following years. He argued that heat and soot from the pipes that were located almost directly beneath the paintings caused permanent damages to the objects of art. Because of this, he requested that the National Board of Building and Planning (which managed the building) would make a serious investigation of how a new system should be designed.[48] In 1925, a committee was appointed to investigate the needs of the museum, with Gauffin as one of the members.

There was also a need for electrification of the museum in order for opening hours to be extended into the evening, making artificial illumination necessary. Gauffin meant that the museum should be able to guarantee the preservation of both art on display, and the art deposited in storage spaces. Good storage facilities with adequate climate was thus necessary, since "these separated objects have the same right as the displayed ones to be preserved for the future—until the day when they perhaps are valued again—the verdict of what is valued and what is unimportant is cast anew by every generation."[49] In these words, Gauffin summarized much of the ethos of modern conservation ideology. If nothing of the collections was to be sacrificed, adequate storage had to be organized for all the collections.

In the annual reports of 1926 and 1927, the Perkins ovens were described as a threat to both staff and collections. A fire inspection resulted in the discovery that wood in the floor had charred due to the heat from the pipes. There was obviously the risk of a fire breaking out, with disastrous consequences for the national treasures. If a pipe would blow and destroy an invaluable piece of art, the loss would be more expensive than a new heating system.

On the other hand, if the museum was heated and illuminated at night, it would have all the chances of becoming a popular institution with an important educational mission. Before 1930, the museum was available for working people only on Sunday afternoons. In the weekdays the museum closed before people left their work. There was a combination of arguments that together motivated the expensive removal of the old oven and the fitting of a modern heating and ventilation system, together with electricity.

---

[48]NM, protocols in museum matters, vol. A2: 59, 11 November, 1924.

[49]*Meddelanden från Nationalmuseum nr 49,* Nationalmuseum, Stockholm 1923, 20.

6 Forced Air Ventilation and Humidity Control

The National Board of Building and Planning gave the task of designing a heating and ventilation system to the ventilation engineer Hugo Theorell, who had long experience from working with buildings of historical value. Among other projects he had fitted central heating in the Royal Palace in 1912. Theorell would have to find a solution that would produce a climate that would be suitable for both collections and humans and that had an acceptable design.[50]

Theorell suggested a solution with hot water being distributed to radiators placed in niches in the galleries, but this was not accepted by the museum. Radiators were accepted in the offices and the small display rooms, but not in the galleries. Visible radiators would not have been aesthetically acceptable in any art gallery of the time. They were considered ugly and would decrease the surfaces available for displaying art. Nationalmuseum needed more space for displays, not less.

In 1929–31 the extensive work of installing heating, ventilation and electricity was carried out. The hallway and the galleries would be heated by air coming from three concealed heating chambers. Air was taken in from the entrance, filtered and heated but not humidified (Fig. 6). Gauffin had worried that panels and frames of wood would run the risk of being dehydrated, but Theorell did not think this was a serious risk. In his final report, Theorell did not mention the importance of RH levels, only considering that temperature was not supposed to vary abruptly.

He was of the opinion—which was common at the time—that temperature should be kept as low as possible where vulnerable objects were kept such as furniture or polychrome wooden objects. It was more risky to use radiating heat than to add already heated air to a room, he said.[51] This argument was developed in an undated pro memoria written by Theorell after his work on Nationalmuseum. The PM has been found in the archive of his firm *Hugo Theorells ingenjörsbyrå*, which is today owned by SWECO and kept at its head office in Stockholm. In it he explores the problems of heating museums, and says that relative humidity and air exchange rate, not temperature, are the most important factors to consider. The importance of keeping a steady and "appropriate" level of RH between 45 and 60% in order to "counter damaging dehydration of the air" is underlined.[52] This leads him to the conclusion that humidified air is the superior way of controlling the indoor climate of museums. He disqualifies the techniques that had been used at Nationalmuseum and other art museums in which staff had been putting bowls of water in the galleries or kept windows open in order to humidify the air. The PM also explains why the air inlets were situated close to the ceilings of galleries:

---

[50]SOU 1931: 8, *Ny värmeledning samt elektrisk belysningsanläggning i Nationalmuseibyggnaden i Stockholm,* Kungl. Byggnadsstyrelsens Meddelanden 3, Stockholm 1931.

[51]Theorell in SOU 1931: 8, 22.

[52]SWECO archive, H. Theorells ingenjörsbyrå, "Uppvärmning och ventilation av museer", 2.

Fig. 6 The new heating and ventilation system installed in 1929–30 drew preheated air from the vestibule and upwards through the building. Inlets were placed in the cornices of the galleries and outlets were integrated in the skylights (*Mouseion* 1934)

> … openings for air intake into a room are arranged so that objects are not directly exposed to the warm flow of air, and it is also preferrable that the movements of the layers of air and air movement around vulnerable objects are kept as small as possible, as the air movement itself boosts the negative consequences, which the character of the air can bring in general.[53]

The PM confirms that Theorell, at least after having done the intervention in Nationalmuseum, was up to date with international research on the subject of indoor climate in museums. It also proves that he considered the control of RH as an important feature of indoor climate in museums.

The effects of a much drier indoor environment quickly became evident to the curator in 1931. Clearly, the last two requirements—human comfort and aesthetics—were considered the most important ones, whereas the climatic needs of the collections were down-played somewhat by the engineer. The indoor temperature was raised since visitors no longer were required to enter the galleries carrying their outer garments. Thus the level of human comfort was raised in several ways and not just by increasing the temperature.

The powerful intake of outdoor air made the environment very dry the first winter. Previously there had been just the natural ventilation of the building, which

[53]SWECO archive, H. Theorells ingenjörsbyrå, "Uppvärmning och ventilation av museer", 4. Author's translation.

of course had been much less efficient than electrical fans. These were working during open hours. Gauffin's fears had proven to be real. Especially panels in the Dutch collection began to show cracks. The new ventilation system needed to be supplemented with humidifiers quickly.[54] In this way, Nationalmuseum was the first museum in Sweden to obtain mechanically humidified air. Apparently, the fitting of humidifiers in 1932 seems to have satisfied the museum for the time being. Issues of humidity do not appear in sources again at Nationalmuseum until the 1950s, when additional humidification was required to meet new norms of indoor climate. Then new requests were made for raising and keeping the RH level steadier than before.

The Nationalmuseum system with heated and humidified air was observed at the 1932 annual meeting of the Scandinavian Association of Museums.[55] In the 1930s with its unmatched period of museum building in Sweden, the issue of climate would be discussed more thoroughly for the first time.[56] The 1932 meeting made it apparent that scientifically based knowledge on how climate affected collections in different ways still was very limited in Scandinavia. It was easy enough to make measurements of temperature and RH levels, but Sigurd Curman who had become Director of Antiquities, wished to know how climate actually affected the different materials of museum objects.[57] Changes in the indoor climate needed to be monitored along with the reactions of collections. A more scientific approach to climate interaction with museum collections would not develop in Sweden until in the 1960s, but in the 1920s the recognition of climate induced problems was for the first time met by a museum in Sweden using humidity control.

7 Humidity Challenging the Building and the Artworks

Instead of lowering the temperature, additional humidity was added to the air from 1932, but the air would still be too dry for much of the furniture, panel paintings and polychrome wooden objects.[58] As a consequence more attention was paid to the relationship between temperature and RH inside the museum, and more humidifiers were installed. Evidence of the dry climate of the building was the return of the collections from their wartime storage space in 1945. They had been stored underground in a very stable climate. Upon their return to the museum, the much drier air had soon damaged furniture. This posed a problem for the museum in the way that objects needed expensive conservation.

[54]RA, KBS, Intendentsbyrån, vol F1A: 124, K. Bildmark's staff letter, 1 September, 1931.

[55]*Skandinaviska museiförbundet. Berättelse över mötet i Lund och Malmö 31 maj–4 juni* 1932, Nordisk Rotogravyr, Stockholm 1932.

[56]Legnér & Geijer (2015).

[57]Curman (1933).

[58]Legnér (2011, 134).

Adding humidifiers to the heating system did not prove to solve the problem of keeping a stable indoor climate, it was a technology that remained in use for the rest of the century. Windows in the galleries were single sash, which caused condensation on the inside of the glass panes. They were not added with secondary glazing for a very long time because of the costs involved. The building was managed by a government agency responsible for the care of state owned properties used for civilian purposes. The agency preferred having higher running costs for heating and ventilation rather than investing in costly improvements of the building. The introduction of moveable humidifiers in the early 1950s was made possible by technological development and was based on the idea that the microclimate could be controlled in every gallery individually. Increased international cooperation put pressure on the museum to better control the indoor climate. This was however not possible to do in the smaller cabinets since these had outer walls with single sash windows. Compartmentalized control of humidity meant that exhibitions had to be organized according to groups of objects: for instance, canvas paintings could not be exhibited together with furniture or paper.[59] This solution proves how pragmatically museum management looked at the issue of indoor climate. When foreign museums demanded a certain climate in order to put artworks on loan, the museum made some efforts to meet the stricter requirements. The in-house collection was instead subjected to seasonal variations in temperature and RH. This management issue could be documented only by using the archive of the museum. Today it has (once again) become possible to display different kinds of objects in the same space.

In other countries it was becoming common to introduce HVAC in national art museums at this time. Increasing car traffic would bring the indoor climate of the museum to the fore in the 1960s. Research on air pollution showed that the building with its leaky windows and entrances functioned as a "chimney" drawing outside air with its particles of dust and tar into the building, making the pollution stick to walls and artworks.[60] Museum management was profoundly sceptical to the idea of sealing the building and introducing full HVAC in order to clean incoming air. The museum was protected by national legislation and according to management it should not be retrofitted. Experience from the Louvre in Paris, which recently had been fitted with HVAC, proved that the system would only collect particles in the air. Sulphur and dioxide coming from industrial outlets and car traffic would not be stopped in this way. These were pollutants damaging paintings, paper and sculpture.

One unpredictable factor affecting indoor climate was visitor attendance. A Rembrandt exhibition in 1956 proved immensely popular as it attracted 290,000 visitors, but despite the humidity added by the bodies additional humidifiers were used in the exhibit space.[61] The agency responsible for caring for the building

---

[59]Riksarkivet K (1953) Byggnadsstyrelsen, Intendentsbyrån, vol F1A: 127. H. Anderdahl's pro memoria "Förslag till befuktningsanordningar…" dated 19 Nov 1953.

[60]Bjurström (1976).

[61]RA, KBS, Intendentsbyrån, vol. F1A: 127, O. Åkerstedt's pro memoria dated 22 September 1955.

meant that the indoor climate should be possible to control without the addition of these extra humidifiers. Museum management was of another opinion and went against the advice of the agency. In an international study of 1960 the museum stated that it attempted to keep an indoor temperature of 18 °C and RH of 50–60%.[62] There is no possibility that the climate was actually kept within this very strict interval, since it was a 19th century building which was far from airtight and did not use an HVAC system. However, in order not to make international loans impossible it was important to state for the records that indoor climate should be kept as stable as possible.

An exhaustive report on the indoor climate problems of Nationalmuseum of 2004 showed that infiltration continued to be seen as a main problem for a very long time. Every hour half of the air volume of the building was exchanged. This uncontrolled inflow of air accounts both for the infiltration of pollutants and for making the air inside very dry in winter. In winter RH could fall well below 40%.[63] In fact, infiltration of outdoor air has been a permanent feature of the building since the 19th century, and this problem has gained increasing attention since the 1950s as the outdoor air became more polluted. Climatization of parts of the building was made possible after the retrofitting and restoration work finished in 2018.

Previously the museum had wished to keep a very strict climate for blockbuster exhibitions on loan from foreign museums, while it had accepted that national collections were displayed in a much less controlled climate. The only way of proving that this actually happened at different occasions has been to visit the archives. It has also become clearer why so little was done for a long time to minimize infiltration of outdoor air. Sensitive objects were not moved to other premises where climate could easier be controlled except for conservation measures. Seen as a management issue, the conservation environment was influenced by a combination of factors of economy, human comfort, preventive conservation and design.

References

Bihang till riksdagsprotokoll (1859–60), vol 11

Borgarståndets riksdagsprotokoll (1859–60). vol 6, pp 363–375 (22 September, 1860)

Bjurström P (1976) Nationalmuseum och Blasieholmsleden. In: Hall T (ed) Nationalmuseum. Byggnaden och dess historia, Stockholm

Bjurström P (1992) Nationalmuseum 1792–1992, Stockholm, p 117

Climatology and conservation in museums (1960). ICOM, Paris

Curman S (1933) Museibyggnader. Byggmästaren. Arkitektupplagan 3(8):33–43

Grandien B (1976) Det Scholanderska fiaskot. In: Hall T (ed) Nationalmuseum. Byggnaden och dess historia. Nationalmuseum, Stockholm, p 66

---

[62] *Climatology and conservation in museums*, ICOM, Paris 1960, 279.

[63] NM, 2004, AB 20 Teknisk förstudie (2004-10-01: 2).

Hawkes D (2012) Architecture and climate. An environmental history of British architecture 1600–2000. Routledge, London, pp 129, 131

Ierley M (1999) The comforts of home. The American House and the Evolution of Modern Convenience. Clarkson Potter, New York, p 127

Laine C (1976) Nationalmuseum och förebilderna. In: Hall T (ed) Nationalmuseum. Byggnaden och dess historia. Stockholm, p 85

Legnér M (2011) On the early history museum environment control: Nationalmuseum and Gripsholm Castle in Sweden, c. 1866–1932. Stud Conserv 56

Legnér M (2015) Conservation versus thermal comfort—conflicting interests? The issue of church heating, Sweden c. 1918–1975. Konsthistorisk Tidskrift 84(3):153–168

Legnér M, Geijer M (2015) Kulturarvet och komforten. Inomhusklimatet och förvaltningen av kulturhistoriska byggnader 1850–1985. Krilon, Klintehamn.

Malmborg B (1941) Nationalmusei byggnad. Ett bidrag till dess tillkomsthistoria. Nationalmusei årsbok, pp 36–37

Manfredi C (2013) La scoperta dell'aqua calda. Nascita e sviluppo dei sistemi di riscaldamento centrale 1777–1877. Politecnico, Milano, pp 52–80

Meddelanden från Nationalmuseum nr 48 (1923). Nationalmuseum, Stockholm

Meddelanden från Nationalmuseum nr 49 (1923), Nationalmuseum. Stockholm

Sander F (1866) Sveriges National-Museum, Stockholm

Sander F (1876) Nationalmuseum. Bidrag till taflegalleriets historia, vol 4. Samson & Wallin, Stockholm, p 1

Sheehan JJ (2000) Museums in the German art world. From the end of the old regime to the rise of modernism. Oxford University Press, Oxford 116–119

Skandinaviska museiförbundet (1932). Berättelse över mötet i Lund och Malmö 31 maj–4 juni 1932, Nordisk Rotogravyr, Stockholm

SOU (1931): 8, Ny värmeledning samt elektrisk belysningsanläggning i Nationalmuseibyggnaden i Stockholm, Kungl. Byggnadsstyrelsens Meddelanden 3, Stockholm

Stålbom G (2010) Varmt och vädrat. VVS-teknik i äldre byggnader. Sveriges VVS Museum—SBUF—VVS Företagen, Stockholm, p 15

Stavenow Å (1972) Nils Månsson Mandelgren, Stockholm, p 122

von Dardel F (1866) Ett besök i Nationalmuseum. Tidskrift för byggnadskonst och ingeniörwetenskap, Stockholm, 2–9

Widén P (2009) *Från kungligt galleri till nationellt museum. Aktörer, praktik och argument i svensk konstmuseal diskurs ca 1814–1845.* Gidlunds, Hedemora, pp 222–223

Willers U (ed) (1977) Hundra år i Humlegården, Stockholm, p 4

Willmert T (1993) Heating methods and their impact on Soane's work: lincoln's inn fields and dulwich picture gallery. J Soc Archit Hist 52:26–58

Asserting Adequacy: The Crescendo of Voices to Determine Daylight Provision for the Modern World

Oriel Prizeman

Abstract This chapter reviews the legacy and formulation of the authority behind emphatic 20th century guidance on daylight design. The degree to which these standards were created in response to the experience of industrial atmospheres in Britain is perhaps unsurprising, however, the extrapolation and determination of standards from precedent has a much less logical path. Whereas the quantification of artificial light was ultimately required to enable its commodification for sale, estimates of adequate quantities of daylight would be based on human experience and methods for its measurement derived in part from seemingly obscure coincidences of legal precedent and anecdotal suggestion.

1 Introduction

It is easy to dismiss the significance of daylight design in all but energy saving or purely aesthetic terms in the context of readily available artificial light, visibly unpolluted skies and sanitised cities with an entirely adult labour force. The breadth of interests whose confluence over a century propelled concerns for light and air to reach their zenith in British post war building programmes can no longer be heard and much of their rhetoric is apparently inapplicable now. Their Victorian sanitarian titles, "Our Homes", "Our Factories", "Our Schools" seem both imperious and paternalistic. The degree of international consensus that did eventually emerge from these has served to foster a rigid approach that has ultimately removed daylight from centre stage as an architectural concern and transformed it into a minimum obligation. It is evident that the requirement for consensus, though based upon progressive research, was repeatedly reasoned by relying on precedent or cross-reference in order to seemingly broaden and strengthen the foundations of their cause. As result, most of the notions, terms and even quantities that are suggested to designers today, stand upon very long legs indeed.

O. Prizeman (✉)
Welsh School of Architecture, Cardiff University, Cardiff, UK
e-mail: prizemano@cardiff.ac.uk

C. Manfredi (ed.), *Addressing the Climate in Modern Age's Construction History*,
https://doi.org/10.1007/978-3-030-04465-7_8

The Regulation of building in Britain was most forcefully introduced following the two greatest events to devastate London; the Fire of London in 1666[1] and the Blitz of 1940. However, between these two events the specific social and environmental circumstances of the nation compelled the issue of daylight to rise to the top of its legal, medical, political and architectural agendas. Rapid industrialisation and urbanisation produced unprecedented atmospheric pollution that, combined with over 150 years of window tax in a generally clouded maritime climate, brought urgency to the requirement for regulatory control.

However, these Victorian ambitions were not played out fully until the composition of British Standard Codes of Practice for Daylighting immediately after the Second World War, in which the necessity of rebuilding was explicitly recognised as an opportunity for improvement. Over a century of discourse to improve living and working conditions had brought daylight to the forefront of the public agenda. By 1932, the quest of "Light and air" in architecture had gained sufficient popular momentum to be of national interest. At its peak, a paper on the orientation of buildings, written collaboratively by the Royal Institute of British Architects and the Illumination Research Committee in 1932 reached the headlines of the Times newspaper.

As measurements were required to secure the application of standards, the history of their formulation is closely related to that of photometry and of the increasing availability of artificial light, which was more readily quantified and compared. Standards were sought for all scales of application, from minimum light for reading to the layout of cities. The subsequent recognition that many of these prescribed quantitative standards were at best misleading and at worst causing unmanageable thermal problems has caused designers and procurers of buildings to retract from ambitions to rely upon daylight and to increase their dependence upon the perceived constancy and increasingly cheap possibility of artificial light.

The 2008 British Standard Code of Practice for Daylighting still justifies the use of daylight in terms of its health benefits, in particular for those unable to control their own environments "a view out-of-doors should be provided irrespective of its quality" (BSI 2008). The recommendation echoes Florence Nightingale's famous nineteenth century charge for patients "to see out of a window from their beds, to see sky and sun-light at least" (Nightingale 1859). However, the current prosaic stipulation that a view should contain three layers of interest; natural distance, man-made middle and foreground (BSI 2008, p. 6), belongs to a primary architectural source; Sir Henry Wooton, in his translation of Vitruvius in 1624, he had written: "For the Sight being a roving and imperious senfe, cannot endure an narrow Circumfcription, but must be gratified both with Extent and Variety" [sic] op cit (Neve 1736).

[1]"By an Act 7 Q. *Anne,* it is directed that after the first of *June* 1709, no Door, Frame, or Window-frame of Wood, to be fixed in any Houfe in *London* and *Westminfter*, and their Liberties shall be fet nearer to the Outside Face of the Wall than 4 Inches" (SIC) NEVE, R. 1736. *The city and country purchaser's and builder's dictionary,* London, B. Sprint [etc.] on windows—a breach was punishable with imprisonment.

The strongest agenda in favour of day lighting today is to conserve energy, although no statutory requirements remain to enforce minimum quantities. Before dismissing the over-compensation of post war standards, it is important to identify the age and recognise the extreme circumstances from which their force was derived. It is also worthwhile to observe the extent to which conventions that were never conceived as statutory instruments have managed to exert a great, if unacknowledged, influence on those that were. Whilst revolutionary and modern principles ostensibly upturned tradition, it is abundantly clear that the arguments used to assert the authority to do so consistently depended on a ladder of precedent that has often been hidden from view.

2 The Health Benefits of Daylight

Evelyn's *Fumifugium*, published in 1661 complained at the phenomenon of "smoke" in England (Evelyn 1661). Industrialised Britain presented a darkened picture of prosperity and heightened the value of the contrasting benefits of daylight, particularly in cities. An article in the Lancet 200 years later described: "London is hemmed in by factories and other places which load the air with dust or smoke, which have a great effect in absorbing the actinic power of the sun's rays" (Black Jones 1898). The limitation of domestic coal burning in British cities was not to be fully addressed until the Clean Air Act of 1959.

Lavoisier's eighteenth century claim that plants and animals required light in order to grow became the key scientific basis upon which the health benefits of daylight were determined (Hesketh 1852; Lavoisier 1790). The simple metaphorical connection between light and dark and good and evil was readily adopted in the rhetoric of eminent scientists; "elements of uncleanliness and corruption which have a vested interest in darkness" (Brewster 1781). By 1909 the idea that morals could even be affected by the visual impact of environmental design was even seriously discussed (Triggs 1909).

The recognition of a need to improve the public urban physical environment rose to a climax within the inequitable class structure of Victorian society that had not only domestic concerns but also assumed imperial authority. 28 years before Engel's "Condition of the Working Class in England in 1844" (Engels) was published, Elsam appealed to landed MPs and Lords to recognise that they had vested interests in securing the welfare of their subjects, who laboured to secure the "pride, honour and independence of the British empire". He detailed how they lived "with no other aperture than a door in a mud wall to let out the smoke" (Elsam 1816).

Florence Nightingale's observations were first published in 1859: "It is a curious thing to observe how almost all patients lie with their faces turned to the light" became internationally famous (Nightingale 1859). In 1866, two rulings on "Rights to the Ancient Lights"[2] and a paper by the head of the Royal Society in Edinburgh, Sir David Brewster, launched a call to action: "it becomes a personal and national duty to construct our dwelling houses, schools, workshops, factories, churches, villages, towns, and cites upon such principles and in such styles of architecture as will allow the life-giving element to have the fullest and the freest entrance" (Brewster 1866). The Royal Sanitary Institute was founded in 1876 "to interpret to the official and general public the methods of applying scientific ideas to the improvement of the environment and to the promotion of individual health" (Nature 1921). Sir Edwin Chadwick was credited as the first to start the modern open space movement with his report "Effect of Public Walks and Gardens on the Health and Morals of the Lower Classes" (Triggs 1909).

By the 1880s, the director of HM Public Works and Buildings noted that it was the "duty" of architects and engineers to follow the principles laid down by "physiologists and medical men" (Galton 1880). Addressing the RIBA, he had stressed that architects should recognise that "sunshine acts as a purifier of the air" (Galton 1874). In the British Medical Journal, William Eassie, pointed to the modernity of sanitary design (Eassie 1874). Daylight was believed to have a sanitising effect in itself (Galton 1880). Texts directed to informing the widest possible audiences concurred on the hygienic effects of sunlight (Murphy 1883).

The fact that window tax, first imposed in Britain in 1696, was not fully repealed until 1851 raised the issue of daylight in parliament. The tax was first associated with health in the Commons in 1815, in an argument to curb the extension of the Window Tax to manufactories, it is significant in relation to daylight guidance today as it identifies workers as the victims of their imposed environments. The Marquis of Lansdowne appealed: "that the tax was most unjust, because it had no connection with the opulence of those on whom it was principally to fall; and that it was inhuman, because of the injurious effect which it must have on the health of those employed in the manufactories" (Lansdowne 1815). The campaign to repeal window tax in Dublin was the first to identify its sanitary effect on the poor: "while the landlords of houses in Dublin, who kept their windows open on the lower floor, were free from the contagion, the poor lodgers above stairs, who closed their windows in order to escape the tax, were most severely afflicted by it" (Shaw 1819). By 1824 the arguments were being applied in favour of the inhabitants of Lambeth and Westminster (Hobhouse 1824).

The campaign struggled as the revenues were said to provide valuable income for war. Three years before the act was finally repealed, the rhetoric of light and air was gaining momentum which was to continue until today: "It was the history of a long series of struggles between the people of Great Britain, and their rulers, in which, he was sorry to say, the health of the people, and the architecture and ventilation of houses in Great Britain, had grievously suffered" (Duncan 1848). "The tax on light" (Murphy 1883) was consciously credited for inspiring the low density planning strategies of Garden City Architects of the early twentieth century:

[2]Dent v Auction Mart Co. (1866) and Butt v Imperial Gas Co. (1866).

"No greater mistake was ever made than the imposition of window-tax—the taxation of the light of heaven. But the degree of efficiency of natural household lighting is dependent on another thing—the *style of architecture* in which the dwelling is built. Let me therefore impress, with all the force I possess, the propriety —nay! Absolute necessity, from the point of view of health—of choosing, in connection with Garden Cities, *only that style* of architecture which *will* lend itself to the provision of *ample window area*" (Sennett 1905).

3 The Precedent of the Right to Light

Unlike many other nations, legal rights to light in England had been derived from precedent drawn from Roman case law. The rights to light and air brought the interests of the individual into conflict with those of the community. "Let them all remember, if they sensibly diminished light to a neighbour's property, they also lessened the free passage of air, and thereby not only risked his health but also their own, and that of the public." (Donaldson 1866, p.187).

In an effort to establish international standards in the late 1920s, Percy Waldram, an architect and fellow of the Royal Sanitary Institute addressed the *Commission Internationale de L'éclairage* [C.I.E.] using the British legal experience of administering rights to light as a means to discuss with authority the issue of public health in relation to daylight (Waldram 1928, p. 478). Waldram used the experience of rights to light both to assert a level of particular British expertise with regard to daylight but also to demonstrate at an international level that these rights could potentially be enshrined in legislation.

4 Asserting the Size of Openings

The first written architectural guidance relevant to daylight were proportional rules for the design of window openings. Following Vitruvius, examples from classical antiquity were used to demonstrate proportional recipes for architectural elements with authority. The processes of translation and abbreviation served to distil observations into rules. A remarkable quantity of these "rules" affected the formulation of modern guidance, albeit indirectly. Initially, "rules" were translated verbosely and allowed a degree of discursive leeway; "all such things are to be increased and diminished at the workman's pleasure" [sic] (Serlio 1611) Book4 Fol 13. Similarly Palladio's qualifications to proportional rules were ironed out or abbreviated in later editions. News from Rome travelled slowly; 123 years after the completion of the Campidogilo, an English "translation" of Vignola featured "A New worke at the Capitole of the invention of Michael Angelo" [sic] (Vignola et al. 1669).

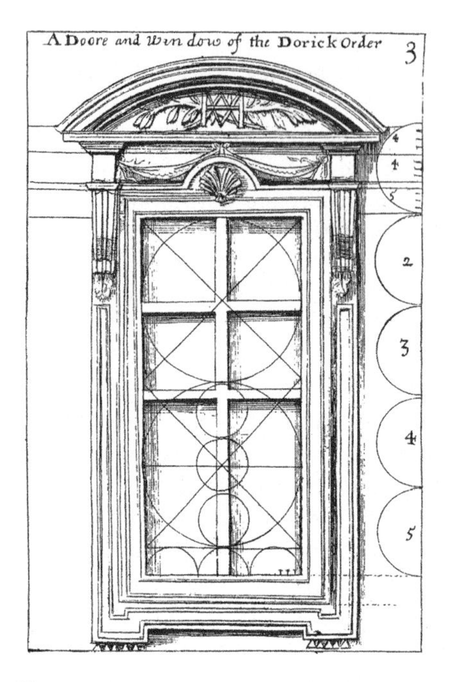

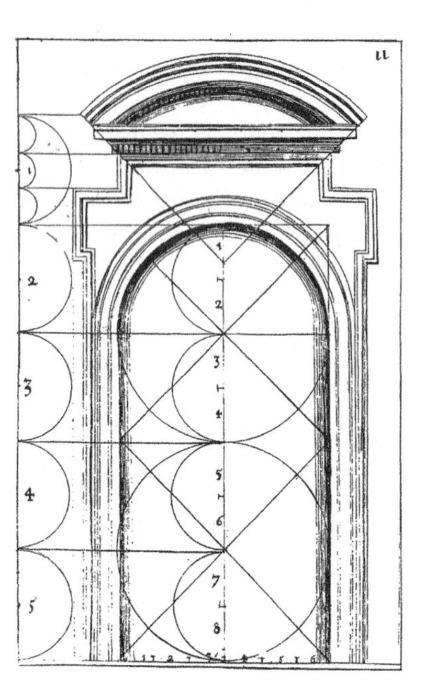

Fig. 1 Doric windows (Pricke 1674)

There was a desire to transmit a scalable catalogue of architectural elements beyond the gentlemen architects to reach "Artificers" directly in the form of pocketbooks. For example, although it was called "The Architect's Store House" John Pricke's book of 1674 was dedicated to tradesmen (Pricke 1674) (Fig. 1). The English translation of Le Clerc's Treatise in 1723 contained dedications to Bricklayers and Carpenters, entreated to "advance it by diligence and imitation" (Le Clerc 1723). The rules were pragmatic and designed to be workable.

There were gradual changes to guidance. Gibbs noted that a compromise must be made with regard to climate "in *England*, we are forced to give Rooms a lower proportion in regard to the coldness of the Climate and the experience of building" [sic] (Gibbs 1732). As early as 1824, one of the first English books wholly concerned with issues of environmental design noted succinctly both the health requirements and the energy issues related to the quantity daylight but also proposed the use of double glazing: "By making the windows double, the loss of heat may be reduced to less than one-third, without sensibly lessening the quality of light" (Tredgold 1824). It may be that Gwilt's observation: "In this country, where the gloom and even darkness of wet, cloudy and foggy seasons so much prevails, it is better to err on the side of too much rather than too little light" (Gwilt 1842) can be identified as contributing to a future affinity for extensive glazing in architecture.

The Palladian guidelines had begun to be questioned in terms of their utility within the British climate early eighteenth century. By the nineteenth century, in his encyclopaedia, Gwilt was concerned to explicitly step back from the application of Palladio's "empirical" rules in England (Gwilt 1842). More boldly, in address to the RIBA in 1874, the director of HM Public Works and Buildings argued that architects should not observe "Italian" rules based on an Italian climate and that the "sanitary principles" of the design of windows should not be "sacrificed to the design of the façade of a building". Arguing that "for sanitary purposes" window heads should be near the ceiling and sills near the floor—he predicted a twentieth century ambition for fully glazed elevations (Galton 1874).

By 1932, the desire to establish standards for fenestration in Britain had reached a climax; the British National Committee (on lighting) in collaboration with its Science Standing Committee and the RIBA circulated "a suggested standard for fenestration to represent modern practice" amongst themselves (C.I.E. 1932). In furthering this aim they distributed their proposal internationally to "representatives of the National Committees of France, Germany, Holland, Hungary, Switzerland and the U.S.A" for comment but received an underwhelming response. The Germans stated that the standard was difficult as there were more things to consider than daylight in design, the Americans more resoundingly rejected the proposal saying "There has never been, as far as I am aware, any interest here in definition of windows of 'ordinary' proportions and of the distances from such windows at which adequate daylight illumination should be reasonably expected … It is felt that such matters call for education rather than for legislation" (C.I.E. 1932). The British preoccupation with daylight design had been fostered under exceptional circumstances explaining their enthusiasm to legislate.

5 The Quest for a Standard

Le Corbusier and Gideon founded the *Congrès Internationaux d'Architecture Moderne* [C.I.A.M] in 1928, meanwhile, the *CIE* met in New York for the second time. Having emerged from the *Commission Internationale de Photométrie*, they were concerned not just with measurement but with design also. Admiring the Illuminating Engineering Society of New York for extending its artificial lighting code to several states, the British committee sought to secure Legislation for minimum standards of daylighting internationally (Britannique 1924).

After some acrimonious discussion it was agreed to base the calculation of a *daylight factor*, the term still used today, then named "daylight ratio", incorporating the Waldram's concept of a uniform sky, avoiding the vagaries of natural light differentiation. The thrust of Waldram's justification was that the system was recognised as "the only existing practical method" having survived 20 years of service as a tool in the British legal system (C.I.E. 1932).

Percy Waldram is credited with the commencement of the "serious calculation of daylighting in an interior in this country" (Collins 1984), he is also latterly questioned for the accuracy of his measurements (Chynoweth 2005). By his own admission, his forceful arguments to establish modern standards internationally ironically relied heavily upon reference to ancient English legal precedent but what is less evident is that even his measurements also incorporated concepts that themselves had long legacies.

Waldram explicitly stated that his proposal of a uniform sky was based upon the "industrial towns in Great Britain" (C.I.E. 1932) to the frustration of several international representatives. Hesketh had first proposed the use of a uniform sky and alluded to the industrial atmosphere of British towns in a paper read at the RIBA in 1852 (Hesketh 1852). His ideas were subsequently championed by his protégée, Professor Kerr.

Even at that point, there was dispute as to the assumption of a uniform sky in London as the basis of calculation, which would be inapplicable in rural situations (Tarn 1869). However, Kerr set an assertive example for Waldram: "The supposition is adopted of an average dull English day; and then, as there is no sun seen, there is an equality of light all around, … If you take this perfectly dull day you admit an equal reflection of light through the atmosphere from horizon to zenith" (Tarn 1869).

The British aspiration for regulations on minimum standards of daylighting to restrict the over densification of cities in the face of an "imminent" need for restrictions in "all countries", was not universally agreed (C.I.E. 1932, p. 213). Most vociferously, the American Committee, represented by Prof Higbie, argued that America was not ready to reach agreements on standard levels of daylight illumination (C.I.E. 1932, pp. 223–4). Despite this a British proposal again succeeded. A plethora of work produced at the National Physical Laboratory in Teddington reflected the British aim to dominate authorship of the subject throughout the 1930s.

The British had chaired the Committee on Daylighting since its foundation in 1927, it was planned in 1939 that the German National committee should take over, however, after the war, it was agreed that this would "prove inconvenient" (Britannique 1948) and the British retained their chairmanship. The 1948 conference was again held at the National Physical Laboratory in Teddington. A marked change in formalisation of the subject of natural illumination clearly took place immediately post war. The research was split into 5 subject areas, each divided for the first time into "daylight" and "sunlight studies".

The authorship of the subject divided by nationality, during the war years displayed a significant acceleration and diversification of interest. Leading at last to resolutions to study "sky characteristics of different regions of the world". By 1948, an artificial sky had been constructed at the Royal Technical University in Stockholm (Britannique 1948). Nevertheless, the initial ambition of Waldram to establish the Daylight factor did succeed in articulating an international term of measurement that is still used today.

After the Second World War, guidance was set out for the first British Standard Codes of Practice. In 1948, the British report had gone so far as to report to the C.I.E. on "good and bad shapes of window bar" (Britannique 1948). Subsequently, the 1949 British Standard Code of Practice for Daylighting formalised the use of graded Daylight Factor Tables related to standard sizes of windows (Britannique 1948). This, together with the post war building programme and the development of mechanised production transformed the fenestration of Britain and introduced legislative minimum standards for daylighting for the first time. The Spartan history of how this authority was determined and derived was transformed by the rapid development of new methods of measurement that enabled daylight to be counted.

6 Measurement and the Translation of Other Languages

Written standards relied upon comparative observation of what already existed. Thus, as Serlio had identified the superior illumination of the Pantheon (Serlio 1611), in the 1850s Hesketh compared numerical proportions of light entering existing buildings against calculations made of the Pantheon (Hesketh 1852).

Trotter's paper was the first to provide a calibrated picture of artificial public illumination of London: "The illumination of the stage of the Lyric Theatre during

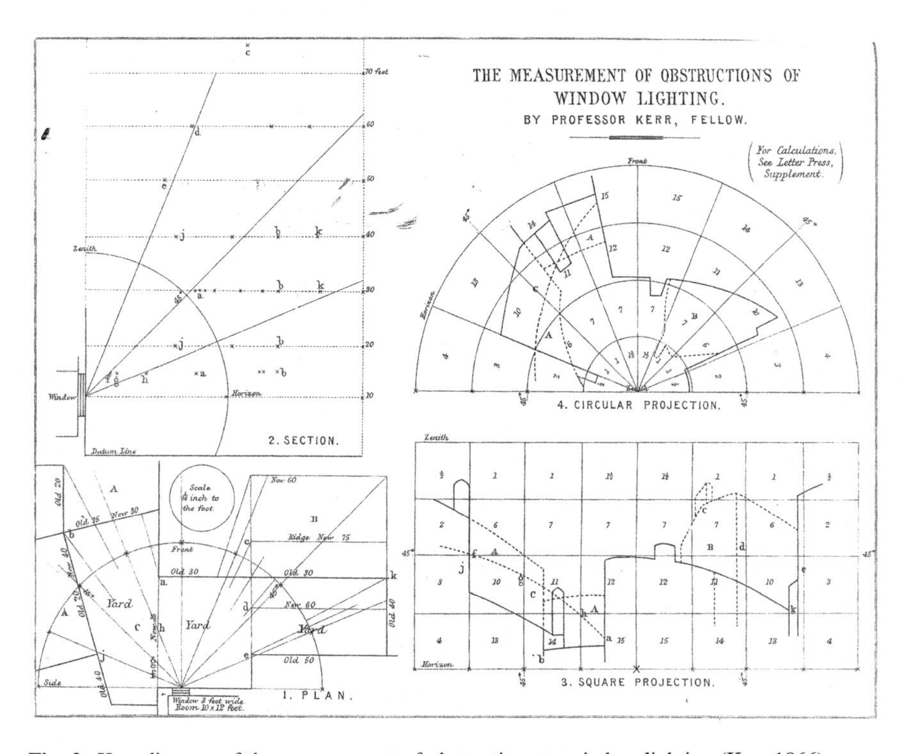

Fig. 2 Kerr diagram of the measurement of obstructions to window lighting (Kerr 1866)

the performance of "La Cigalle" was 3.8 candle-feet without the arc or lime-lights … The illumination in the trains on the Metropolitan and District Railways was measured on many occasions, and varied from 0.3 to 0.9 cd foot, the photometer being held breast-high" (Trotter 1857).

A range of studies into visual comfort emerged in the early twentieth century, encouraging measurements to be converted swiftly to prescriptions. Matthew Luckiesh, Director of the General Electric Lighting Research Laboratory dominated the American work. The British criticised the close relationship of these papers to artificial lighting. Indeed, Luckiesh's studies into desired illumination intensities for reading were attacked by a psychologist for "prescribing foot-candle standards of illumination; and … to help persuade consumers that a certain level of light intensity is necessary" (Tinker 1941). The divergence between the relatively simple terms required to specify artificial light as opposed to the more complex calculations required to predict adequate levels of natural light remain as stumbling blocks for designers today (Fig. 2).

The fact that a loss had to be modified in cases of rights to light, stimulated the development of standard methods of measurement for daylight in Britain. Architects were urged to equip themselves with the tools of "scientific proof" in

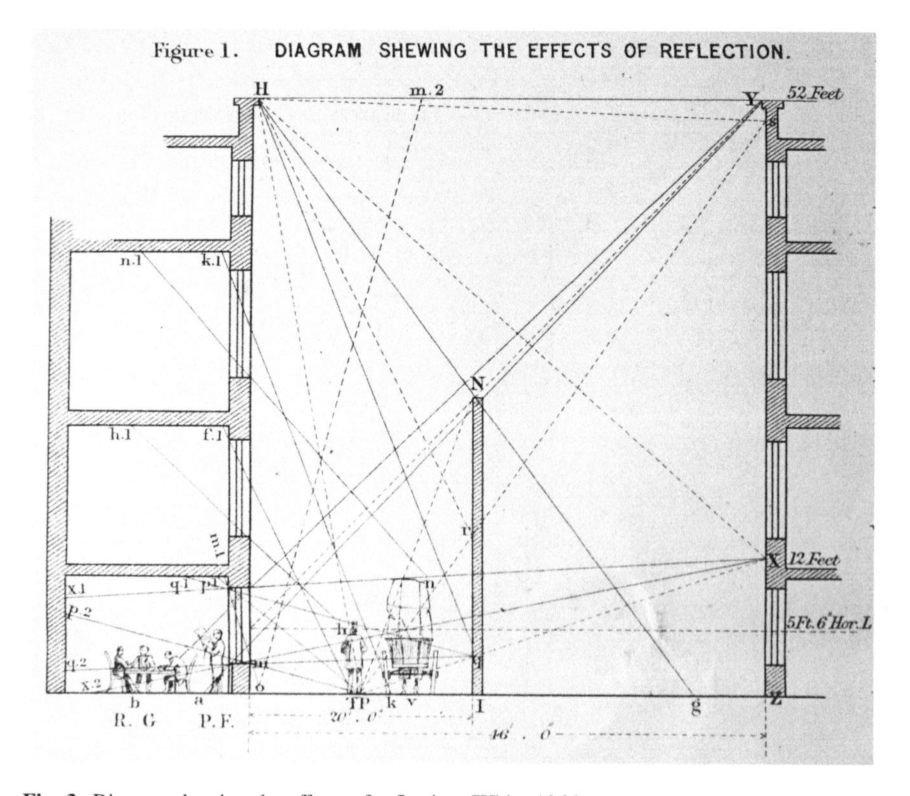

Fig. 3 Diagram showing the effects of reflection (White 1866)

cases of rights to light (Kerr 1866). Kerr proposed a method of proportional measurement based on Hesketh's work, from which he suggested standards of "necessary lighting" could be agreed. Most importantly, he asserted the principle of a standard of light described as a percentage of the sky surface. The paper stimulated fierce debate, both as to points of law as well as to standards of measurement and their means of expression (Kerr 1866) (Fig. 2).

Later that year, another RIBA fellow attempted to challenge Kerr's assertion that reflected light was of negligible importance (White 1866) (Fig 3). Kerr argued in defence of his claim "I cannot measure light and air like wheat in a bushel, or water in a cistern, but I can get a *proportionate* measurement of light abstracted; I can calculate a percentage" (Donaldson 1866). This concept was the seed of the daylight factor.

7 The Daylight Factor

The desire to provide architectural formulae to determine requisite quantities of daylight illumination made slow progress through the scholarly adaptation of Palladian rules during the seventeenth and eighteenth century. What is less obvious is how twentieth century principles were closely related to these early lessons.

Salmon's "Palladio Londinensis" used the Baths of Titus in Rome as an exemplar of proportions of windows in relation to light (Palladio 1728). Hoppus added to the collection using sciagraphy to illustrate the volume of the room with shadows (Hoppus 1737). Gibbs developed elaborations (Gibbs 1732). In the nineteenth century, Gwilt's encyclopaedia related window area to volume for the first time, quoting Robert Morris. "Let the magnitude of any room be given, and one of those proportions I have proposed to be made use of, or any other; multiply the length and breadth of the room together, and that product multiply by the height, and the square-root of that sum will be the area of superficial content in feet, & c., of the light required" (Morris 1734). The formula endured through Hesketh as 10 a ratio of square foot of glass to 100 ft^3 of room.

HM Director of Public Works proposed Morris's same ratio to the RIBA in 1874 (Galton 1874). A window area of 14% had been cited in the building regulations (Metropolitan Building Act 1855). The Metropolitan Building act of 1905 again set the area of windows at one tenth of that of "Every habitable room" (Dicksee 1908).

Trotter addressed the issue of the photometric measurement of the Daylight illumination of interiors in 1895 (Croghan 1963) and again referred to London. Using a model which assumed that all "illumination were produced by light from a uniformly grey sky". He justified this saying, "On a considerable number of days in the year, the sky of London, and to a less degree, in other parts of England, is of a more or less uniform grey." He modified a photometer to imitate a hemispherical cover with stops varying in diameter from 1/200th to 1/5000th and took measurements outside and inside to provide a coefficient for the room. Waldram took up his system in order to make a series of comparative measurements of public rooms (Trotter 1910).

Coinciding with the outbreak of the First World War, the Illuminating Engineer reported that the "daylight ratio" had gained momentum as an internationally agreed concept between German and British researchers, Pleier had established a formula for calculating an "average direct illumination at any point in a given building" relative to outside sky brightness. Weber used this for the purpose of establishing minimum standards for schools, Waldram concurred with his result (Gaster 1914). A government technical paper, established Waldram's "sill ratio" or daylight factor as a "recognised criterion of interior lighting" (DSI 1927). By 1949, the depth of daylight penetration into the room was illustrated relative to the working plane as a British Standard.

The British aim to achieve an "international agreement" on the form and expression of a "Daylight Ratio" stumbled in negotiation over the concept of "skies of uniform brightness" The standard notion was seen as irrelevant in other countries. "in this country (USA) we have much more daylight than they have in England" (C.I.E. and Macintyre 1928). Indicatively, each team proposed different quantities in different units. The British technical team had suggested a standard of 500/π international foot candles per square foot, the Germans, 3000/π hefner candles per square metre and the Americans 682/π international foot candles per square foot. Finally, it was agreed that the standard sky brightness be set at 5000/π international candles per square meter, so using the German unit with the British quantity (+7.5%). With somewhat less difficulty it was agreed that the universal height for taking measurements be set at 85 cm "for ordinary sedentary work at tables."

8 Standards Applied to Tasks

Whilst studies concentrated on the orientation of hospitals in relation to sunlight, school buildings initiated unprecedented international concern for daylight. A German ophthalmologist, Herman Cohn had argued that poor lighting in schools caused myopia. Cohn's work was widely published (Cohn and Turnbull 1886). Using a measure of "solid angle of daylight" he compared standards of glazing ratios internationally. The adequate illumination could then be expressed as a percentage of the total possible illumination in each case (Weber 1911).

Standards for school illumination were the first to be enforced in British statute. The IES felt that that the 1907 regulations from the Board of Education in the UK were inadequate and urged a functional approach to school design: "Windows should never be provided for the sake of external effect" (Gaster 1914). In 1924, the CIE set international standards for the illumination of specific rooms in schools for the first time, libraries at 50 lx and 80 lx for art and craft rooms (C.I.E. 1924). They agreed the location of windows to he left, that ceilings should be white and the wall behind the blackboard should be dark. At the 1932 conference, the British committee referred to a 1914 report that recommended minimum daylight ratios for schoolchildren in Britain and Germany at 0.5%.

The most demanding requirements for design with daylight ever established were set for schools in Britain after the Second World War when these figures were radically increased. Not only this, but they had been written into legislation. The first British Standards quadrupled the 1914 standard to demand a minimum sky factor of 2% across the working area. (B.S.I; Codes of Practice Committee for civil engineering 1949). In addition it stated that "sunlight should be able to enter ordinary classrooms each day for not less than 2 h in mid-winter". The 1944 Education Act recommended that a 5% sky factor level should be attained if possible, although the standard enforced stated "At each desk or place of work in every teaching room in every school or department the Daylight Factor shall not be less than 2 per cent" (Britannique 1948).

The issue of daylight in factories was challenged in the same year as Cohn's work was published but with less dramatic responses. Thwaite argued that Britain should compete with international standards (Thwaite 1882). He catalogued the number of deaths per capita by trade in France, Germany and England and derided the transformation of working environments from well ventilated domestic scaled workshops to factories where "rooms were exceedingly low and deficiently lighted" (Thwaite 1882). He argued that the structural mass of the building should be minimised in order to admit the maximum quantity of light and that by increasing the height of rooms they would admit more light and so lengthen the working day in winter by 30–60 min thereby increasing profits (Thwaite 1882) (Fig. 4).

He designed a model mill which accommodated heating and ventilation into structural columns and used pre-stressed concrete lintels to maximise glazing. The efficient design was a vision of modernity that intended to combine sanitary benefits with commercial appeal (Thwaite 1882).

The increasing prevalence of clerical work in the early twentieth century became a focus for research and observations (DSI 1930). In 1931, in search of a standard, a "jury" of "three architects, two illuminating engineers, one accommodation officer for HM office of Works and one engineer" visited twenty offices in Whitehall and compared their subjective impressions against contour maps of photometrically measured daylight factors in each office. A "grumble point" was abstracted into standard sections of rooms in order to represent the impact of obstructions. The iso contour method of representing daylight factors was then adopted at the CIE in 1932 (Britannique 1932). The paper monitoring their responses concluded that a daylight factor of 0.2% was intolerable for office work (DSI 1931). By 1948, a minimum standard of 1% sky factor was recommended; half that stipulated for schools. The daylit zone was further required to penetrate a zone 12 ft. into the rooms, thereby making Thwaite's nineteenth century suggestion to heighten ceilings in factories into a twentieth century stipulation (Britannique 1948).

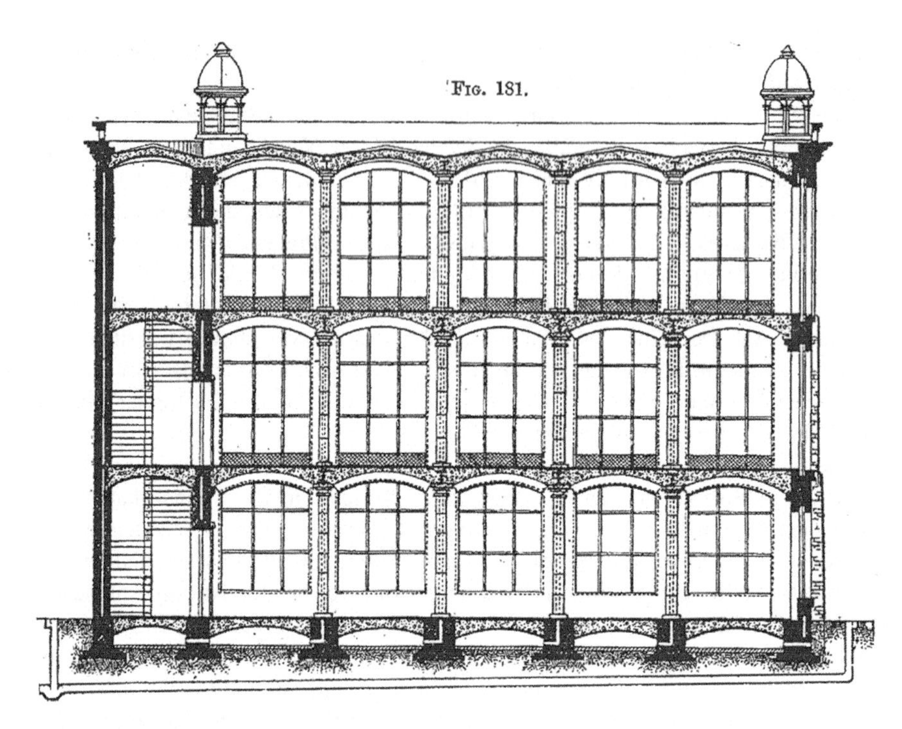

Fig. 4 A model industrial structure (Thwaite 1882, pp 264–265)

9 Rights to Light Enforced at an Urban Scale: Low Density Planning

The perceived threat of over densification, particularly the example of nineteenth century New York stirred anxiety, Forbes Winslow related to British audiences reports of insanitary tenant housing there (Winslow 1867). Waldram illustrated the threat of the "inevitable" invasion of American skyscrapers (Waldram 1928). In confirmation of Waldram's fears, the American Committee of the Illuminating Engineering Society on Daylight commented that it did not consider "it will be possible to advocate any rules or regulations for the amount of light to be admitted to buildings" Arguing that Waldram's proposals to regulate daylight could only be applied to country villages, not cities, they stated with reasoning that was to become widely shared: "Men can exist in rooms without daylight and yet be fairly healthy. Daylight is necessary or light is necessary, to see the dirt in the corners of the room, and to keep the place clean. However, for that purpose artificial light can be used" (C.I.E. and Macintyre 1928).

The regulation of private life was adopted as the means to protect public health in Victorian England. Between 1862 and 1894, the London Building Acts had incrementally reduced the permitted heights of neighbouring buildings with respect to light several times. In 1892 Guidance for the spacing of buildings "inhabited by persons of the working class" in Britain restricted the permitted height of houses to 63½° at the rear (Dicksee 1908) (Fig. 5).

It had long been acknowledged that Pompeian houses turned their backs to the street and admitted light to their cores. Neve referred to "terracing" to admit light in Greek and Roman courtyards (Neve 1736) on building. A paper presented by H.M Office of Works to the C.I.E. in 1928 established that a single courtyard plan for the erection of a large building on an urban island site offered the optimum percentage of direct light area to total floor area in offices (West 1928). Significantly, Raymond Unwin, architect and champion of Ebenezer Howard's Garden city low density architecture, who had used daylight diagrams to set out the spacing of buildings in Garden Cities in 1909 (Unwin 1909), was a member of the Secretariat Committee of the 1928 C.I.E. representing both the Ministry of Health and as a co-author of the universal term for a daylight ratio (C.I.E. and Macintyre 1928).

Diagram III.—Courts within a Building—Closed on all Sides. Sec. 45 (pp. 63 and 64)

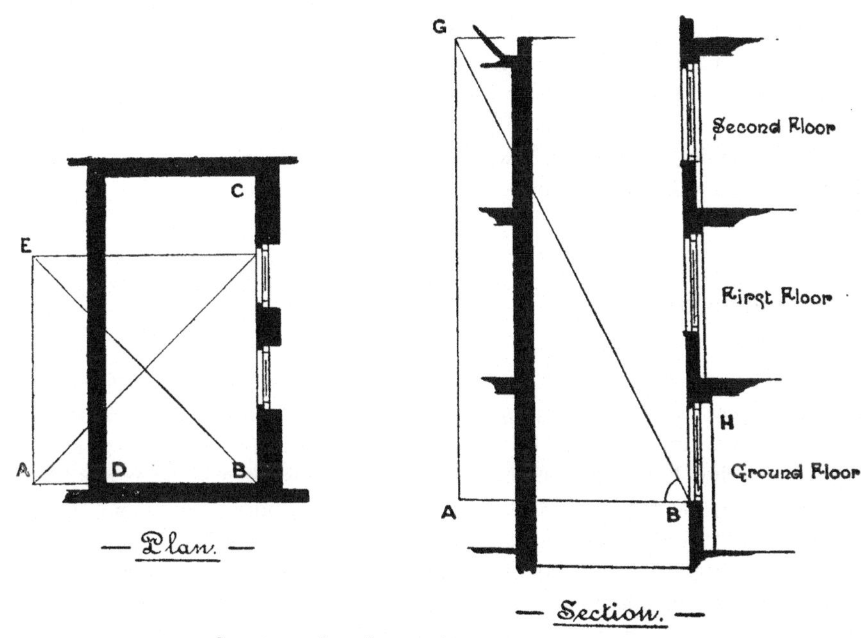

D, B, C. Court enclosed on all sides, the length B, C being not greater than twice the width B, D.

E, A, B. Square equal in area to court, D, B, C.

H. Window on ground floor lighting habitable room from court

G, A. Height of wall opposite to windows, H, measured from level of sill, B, must not be greater than twice A, B.

The angle A, B, G equals 63½ degrees.

Fig. 5 Building regulations for light in enclosed courts (Dicksee 1908, p. 62)

As early as 1942, the Building Research Station in connection with the Lighting of Buildings Committee had produced a report on the definition of good planning practice regarding daylight proposed with the aid of a Waldram diagram that blocks be set at 45° to the street frontage and 90° to one another in order that "Substantially larger areas of sky thus become visible from the working plane at greater depths inside the rooms and wall surfaces receive sufficient light to effect a very useful reduction of brightness ratios" (Britannique 1948). The work went on to promote the use of cruciform, "T" and "L" plans.

A year after the Congrès Internationaux d'Architecture Moderne met to discuss the Reconstruction of Cities in Bridgewater, Somerset, the British Lighting Committee at the C.I. E in 1948 noted "the war has resulted, of course, in considerable open spaces in urban districts, … an administrative basis for the control in terms of daylighting has been provided". The new Ministry for Town and Country Planning (Allen and Crompton 1947) had published a manual on urban redevelopment giving "official guidance" with regard to daylighting (Planning 1948). Together with guidance notes (Pound 1947) and the Code for the rebuilding of London (Holden 1947), authority with respect to daylight design in Britain had ostensibly been achieved. The scope of recommendations was thorough and included both sunlight penetration and daylight factors that were later incorporated into the British Standard Code For Daylighting in 1949. The C.I.E. in 1948 went so far as to propose a British Standard to state that "a minimum of 1 h of sunshine during not less than ten months of the year" should penetrate the living rooms of Britain (Britannique 1948), this however, proved to be an ambition that was beyond the power of legislation to enforce.

10 Conclusion

The broad based enthusiasm for design with daylight in Britain reached its zenith in the aims of post war rebuilding programmes, whose guidance and codes of practice affected architectural intentions from specific requirements for reading to principles of town planning. It is clear that the majority of the terms and parameters that frame guidance today can be traced directly to nineteenth century ideas, if not earlier. Whilst historically the purpose or intention of guidance may have changed from defending the rights of the overshadowed individual to ensuring common hygienic standards for schoolchildren to concerns for energy conservation, the measures used to define these requirements have developed from core precedents. The impact of guidance has varied historically from aesthetic design advice, to legal requirements, yet it is always reliant upon the use of established common terms for its expression, the first measured from the perceived wisdom of classical architects, the second derived from Roman Law. Whilst there is no current statutory requirement for daylight design, the terms, aims and criteria for Daylight guidance in Britain today can largely be traced directly to a legacy of responses to the findings of medical research, the slow battle to end taxation on windows and legal precedent. These arguments have, at their rhetorical height elevated the status of daylight design to that of a human right requiring legislative standards to protect it. In theory, once light had become quantified as a legal commodity, it could be projected into proposals for the future. The basis of the common terms devised to express international standards for measuring daylight was initially justified specifically by their durability in British legal practice. Yet these resultant terms were not derived or authored from the experience of the design process, nor by architects and as such have at times been applied to principles of design in cumbersome terms reliant on testing models. By contrast, the relatively simple quantification of artificial light, easily specified and predicted has further served to threaten the competence of daylight design.

References

Allen W, Crompton D (1947) A form of control of building development in terms of daylighting. R.I.B.A. J

Black Jones DW (1898) The measurement of sunshine at our health resorts. The Lancet 256–259

Brewster D, SIR, 1781–1868 (1866) Address delivered at the opening session 1866–67 of the Royal Society of Edinburgh. In: Proceedings of the Royal Society of Edinburgh for 1866–67, Edinburgh

Britannique CN (1924) The field for International Agreement and Standardisation in Illumination. In: C.I.E. (ed) Commission Internationale de L'éclairage En succession a la Commission Internationale de Photométrie Sixième Session. CUP, Genève

Britannique CN (1932) Recueil des travaux Et Compte rendu des séances: Comite D'Etudes Sur L'Eclairage Diurne Rapport Du Comite Secretariat (Comite Britannique). In C.I.E. (ed) Commission Internationale de L'éclairage En succession a la Commission Internationale de Photométrie Huitième Session. National Physical Laboratory, Teddington, U.K

Britannique CN (1948) Recueil des travaux Et Compte rendu des séances: Comite D'Etudes Sur L'Eclairage Diurne Rapport Du Comite Secretariat (Comite Britannique). In: C.I.E. (ed) Commission Internationale de L'éclairage En succession a la Commission Internationale de Photométrie Onzième Session. National Physical Laboratory, Teddington, UK

BSI (2008) BS 8206-2:2008 Lighting for buildings. In: BSI (ed) Code of practice for daylighting

B.S.I; Codes of Practice Committee for Civil Engineering, P. W. A. B. (1949) In: Works, M. O. (ed) British Standard Code of Practice CP 3-Chapter 1(A) (1949) Code of Functional Requirements of Buildings Chapter 1(A) Daylight (Dwellings and Schools)

Chynoweth PUOS, Salford UK (2005) Progressing the rights to light debate, Part 2: The grumble point revisited. Struct Survey 23:251–264

C.I.E. (1924) Rapport de Comite d'études sur l'éclairage dans les usines et dans les écoles. In: ÉCOLES, C. D. É. S. L. É. D. L. U. E. D. L. (ed) Commission Internationale de L'éclairage En succession a la Commission Internationale de Photométrie Sixième Session. Genève juillet 1924: CUP

C.I.E. (1932) Recueil des travaux Et Compte rendu des séances: Comite D'Etudes Sur L'Eclairage Diurne Rapport Du Comite Secretariat (Comite Britannique). Commission Internationale de L'éclairage En succession a la Commission Internationale de Photométrie Huitième Session. National Physical Laboratory, Teddington

C.I.E., Macintyre JA, Munby A, Unwin R, Waldram PJ, Walsh JWT, West JG, Buckley H (1928) Recueil des travaux Et Compte rendu des séances: Comite D'Etudes Sur L'Eclairage Diurne Rapport Du Comite Secretariat (Comite Britannique). Commission Internationale de

L'éclairage En succession a la Commission Internationale de Photométrie Septième Session. Sarnac Inn, N.Y. Septembre 1928
Cohn HL, Turnbull WP (1886) In: Turnbull WP (ed) The hygiene of the eye in schools By Hermann Cohn … An English translation. Simpkin, London
Collins JBBE, MIEE, CENG, FCIBS (1984) The development of daylighting—a British view. Light Res Technol 16:155–170
Croghan D (1963) The measurement of daylight and its effect on the design of buildings and layout particularly in housing development David Croghan. Ph.D., University of Cambridge
Dicksee BJ (1908) In: Dicksee B (ed) The London building acts, 1894 to 1908: (57 & 58 Victoria, Cap. CCXIII, 61 & 62 Victoria, Cap. CXXXVII, 5 Edwardus VII, Cap. CCIX, 8 Edwardus VII, Cap. CVII); with copious index, notes, cross references, legal decisions and diagrams, also the bylaws and regulations. E. Stanford, London
DSI IG, Macintyre W (1927) In: RESEARCH, D. O. S. A. I. (ed) Illumination research technical paper No 7; Penetration of daylight and sunlight into buildings. HMSO, London
DSI IG, Macintyre W (1931) In: RESEARCH, D. O. S. A. I. (ed) HMSO, London Illumination research technical paper No. 12. The daylight illumination required in offices
Donaldson P (1866) On the practise of architects and the law of the land in respect to easements of light and air. RIBA Trans
DSI (1930) Illumination research technical paper No 10. The effect of distribution and colour on the suitability of lighting for clerical work. In: Research, D. O. S. A. I. (ed) HMSO, London
Duncan V (1848) In: Hansard (ed) The window duties
Eassie W (1874) Sanitary arrangements for dwellings: intended for the use of officers of health, architects, builders, and householders. Smith, Elder, London
Elsam R (1816) Hints for improving the condition of the peasantry in all parts of the United Kingdom, by promoting comfort in their habitations … with … characteristic designs for cottages; … to which are added explanations and estimates made accordingly. Farnborough, 1971
Engels F (1844) Condition of the working class England 1844
Evelyn J (1661) Fumifugium, or the Inconveniencie of the Aer and Smoak of London Dissipated Together With some Remedies humbly Proposed, London
Galton D (1874) Some of the sanitary aspects of house construction. RIBA Trans
Galton D (1880) Observations on the construction of healthy dwellings namely houses, hospitals, barracks, asylums, ETC. Oxford Clarendon Press, Oxford
Gaster LEA (1914) Interim report on the daylight illumination of schools. In: Gaster LEA (ed) Illuminating engineering society London. Illuminating Engineering Society London with the approval of the Council and of the delegates of the Associations represented on the Joint Committee, London, appointed in 1911
Gibbs J, 1682–1754 (1732) Rules for drawing the several parts of architecture,: in a more exact and easy manner … by which all fractions, in dividing the principal members and their parts, are avoided. W. Bowyer for the author, London
Gwilt J, 1784–1863 (1842) An encyclopaedia of architecture : historical, theoretical, and practical/ by Joseph Gwilt; illustrated with more than one thousand engravings on wood by R. Branston, from drawings by John Sebastian Gwilt, London Longman, Brown, Green, and Longmans
Hesketh R (1852) On the admission of daylight into buildings, particularly in the narrow and confined localities of towns. Papers Read at the RIBA 1852–3
Hobhouse M (1824) In: Hansard (ed) Repeal of the window tax
Holden CHAH, CH (1947) Report on reconstruction in the city of London. London
Hoppus EE, D. 1739 (1737) The gentleman's and builder's repository. Gregg, Farnborough, 1969
Kerr P (1866) Remarks on the evidence of architects concerning the obstruction of ancient lights, and on the practice of proof by measurement; with reference to recent cases in the courts of equity. RIBA Trans 1865–6
Lansdowne MO (1815) In: Hansard (ed) 1815 Tax on windows of manufactories
Lavoisier AL, 1743–1794. Kerr R, Mckie D (1790) Elements of chemistry: in a new systematic order, containing all the modern discoveries. Edinburgh: printed for William Creech, and sold

in London by G. G. and J. J. Robinsons, 1790. Facsimile reprint of original (1790) Kerr translation. Published by Courier Dover Publications, 1984
Le Clerc SB, 1637–1714 (1723) Traitel̀ d'architecture avec des remarques et des observations. English: A treatise of architecture, with remarks and observations. Necessary for young people, who wou'd apply themselves to that noble art. By Seb. Le Clerc, … London: printed: and sold by W. Taylor: W. and J. Innys; J. Senex, and J. Osborne, 1723–24
Morris R, 1701–1754 (1734) Lectures on architecture: consisting of rules founded upon harmonick and arithmetical proportions in building. Design'd as an agreeable entertainment for gentlemen: and more particularly useful to all who make architecture, or the polite arts, their study. Printed for J. Brindley, London
Murphy SF (1883) Our homes and how to make them healthy [Shirley Forster Murphy]. Cassell, London
Nature (1921) Royal Sanitary Institute: Folkstone Congress. Nature 107:567–568
Neve R (1736) The city and country purchaser's and builder's dictionary. London, B. Sprint [etc.]
Nightingale F (1859) Notes on nursing what it is, and what it is not. Harrison, Bookseller to the Queen, London
Palladio A, 1508–1580. Campbell C (1728) I quattro libri dell'architettura. Book 1. English: Andrea Palladio's five orders of architecture. With his treatises of pedestals, galleries, … Together with his observations and preparations for building; and his errors and abuses in architecture. Faithfully translated, and all the plates exactly copied from the first Italian edition printed in Venice 1570. Revised by Colen Campbell, … To which are added, five curious plates … invented by Mr. Campbell. Printed for S. Harding, London, 1729
Planning, M. O. T. A. C (1948) In: Planning, M. O. T. A. C. (ed) Redevelopment of central areas, a manual published by the Ministry of Town and Country Planning, London
Pound GT (1947) In: Institute, T. P. (ed) Space between buildings for daylight
Pricke R (1674) The architects store-house: being a collection of several designs of frontispieces, doors etc. Likewise a representation of the five columns of architecture … containing 50 copper-plate-prints Facsimile reprint of the London 1674 ed. Gregg, Farnborough, 1967
Sennett AR (1905) Garden cities in theory and practice, vol 1
Serlio S, 1475–1554 (1611) The first[-fift] booke of architecture, made by Sebastian Serly, entreating of geometrie. Translated out of Italian into Dutch, and out of Dutch into English. First booke of architecture, Peake, Robert, Sir, 1592?–1667, tr., publisher. Stafford S, fl. 1596–1626, printer. Snodham T, d. 1625, printer. London: Printed [by Simon Stafford and Thomas Snodham] for Robert Peake, and are to be sold at his shop near Holborne conduit, next to the Sunne Tauerne, Anno Dom. 1611. (Printed at London: by Simon Stafford [and Thomas Snodham], 1611.)
Shaw MOD (1819) In: Hansard (ed) Motion for the repeal of the window tax in Ireland
Tarn EW (1869) On the admeasurement of sky in cases of light. RIBA Trans
Thwaite BHBH (1882) Our factories, workshops, and warehouses: their sanitary and fire-resisting arrangements. E. and F. N. Spon, London
Tinker MA (1941) Effect of visual adaptation upon intensity of light preferred for reading. Am J Psychol 54:559–563
Tredgold T (1824) Principles of warming and ventilating public buildings, dwelling houses, manufactories, hospitals, hot-houses, conservatories, & c.; and of constructing fire-places, boilers, steam apparatus, grates, and drying rooms; with illustrations experimental, scientific, and practical. Printed for J. Taylor, London
Triggs HIHI, 1876–1923 (1909) Town planning, past, present and possible. Methuen & co, London
Trotter AP, 1857–1910 (1911) Illumination, its distribution and measurement, by Alexander Pelham Trotter. Macmillan and co., limited, London
Unwin R (1909) Town planning in practice: an introduction to the art of designing cities and suburbs. T Fisher Unwin, London
Vignola, Michelangelo Buonarroti, Leeke J (1669) Regola delli cinque ordini d'architettura. English The regular architect, or, The general rule of the five orders of architecture of M.

Giacomo Barozzio da Vignola [microform]: with a new addition of Michael Angelo Buonaroti/ rendered into English from the original Italian, and explained by John Leeke … for all ingenious persons that are concerned in the famous art of building. The general rule of the five orders of architecture. The regular architect., London : Printed for William Sherwin, and are to be sold by … Rowland Reynolds …, 1669

Waldram PJ (1928) Daylight and public health. In: C.I.E. (ed) Commission Internationale de L'éclairage En succession a la Commission Internationale de Photométrie Septième Session. Sarnac Inn, N.Y. Septembre 1928

Weber PL (1911) Some replies to queries on daylight illumination. Illum Eng 1911:34–37

West JG (1928) Daylight illumination in respect to a typical building site in a large town. Commission Internationale de L'éclairage En succession a la Commission Internationale de Photométrie Septième Session. Sarnac Inn, N.Y., Septembre 1928

White W (1866) On the measurement of the observation of ancient lights: further investigation. RIBA Trans

Winslow F, 1810–1874 (1867) Light: its influence on life and health. Longmans, Green, Reader & Dyer, London

The Houses of Parliament and Reid's Inquiries into User Perception

Henrik Schoenefeldt

... sometimes Members come to me, and say the House is very hot, or very cold; I look at the thermometer, and see if so, for different people have different feelings with regard to temperature. People come in very hot, and say, "How cold the House strikes;" and another man says "I have been sitting here half an hour, and I am in fever:" and if I see the thermometers are too high or too low, I give directions accordingly [Select Committee on Lighting the House, Report from the Select Committee on Lighting the House (HC 1839, 501).].

William Gosset, Sergeant-at-Arms, 1839.

Abstract In the first half of the 19th-century building services were the subject of extensive experimental inquiries. In addition to technical trails these inquiries also covered research into human factors, such as the perception of indoor climates and air quality. This chapter investigates how studies into the nature of thermal comfort, undertaken under the direction of the physician David Boswell Reid, had informed the design of the environmental control system in the debating chamber of the UK Houses of Commons. The studies included experiments with test chambers, undertaken in Reid's Laboratory between 1834 and 1836, and empirical observations inside the two temporary debating chambers for the Houses of Commons (1836–51) and Houses of Lords (1838–47). The debating chambers enabled Reid to test and refine his concepts under real-life conditions, involving politicians directly in the process of evaluating and improving the indoor climate.

In 1839 the Sergeant-at-Arms William Gosset had a long interview with a Select Committee appointed to review the effectiveness of the stack ventilation system in the Temporary Houses of Commons. This had been introduced by the physician and chemist David Bothwell Reid in the winter of 1836 and over the following two years the Sergeant-at-Arms had been responsible for gathering and processing oral feedback MPs on their perception of the level of thermal comfort and air quality within the chamber. At time he gave orders to the attendants of the ventilation to

H. Schoenefeldt (✉)
University of Kent School of Architecture, Canterbury, UK
e-mail: H.Schoenefeldt@kent.ac.uk

C. Manfredi (ed.), *Addressing the Climate in Modern Age's Construction History*,
https://doi.org/10.1007/978-3-030-04465-7_9

make adjustments. In his interview Gosset highlighted the management of the ventilation was confronted with the challenge of reconciling the results of measurements with the subjective feedback from MPs. This concerns about user-perception represented a central theme in Reid's work in the field of ventilation. In addition to scientific research on the physiological effect of air quality and climate he used experimental studies to underpin the development of his ventilation scheme for the Houses of Parliament with empirical evidence. These experiments began with a test chamber in Edinburgh (1836), and was followed by the Temporary Houses of Commons (1836–51) and the Temporary Houses of Lords (1838–47). The temporary debating chambers allowed him to test and refine principles under real-life conditions over a period of fifteen years and directly fed into the development of a highly sophisticated climate control system realized in the Permanent House of Commons (1846–52). This paper provides a brief overview of the role of user perception in the development of Reid ventilation system for the Palace of Westminster. User-perception was used as a performance indicator in the day-to-day management of the ventilation, but also it was also a major design factor underlying the development of the ventilation system for the Permanent Houses of Commons.

1 Empirical Approaches

The working methods that were deployed in the development of the Houses of Parliament's ventilation system built on scientific working methods that Reid had developed in the early 1830s to study thermal comfort and air quality. In a lecture entitled *Progress of Architecture*, Reid explained that these experiments involved human participants, who on their perception of the climatic and atmospheric conditions inside experimental chambers (Reid 1856). The aim of these studies was to define new standard ventilation rates, which addressed problems of air quality as well as thermal comfort. He criticized the typical levels of air supply in public building for being too low, resulting in atmospheric conditions that were unpleasant and made occupants feel physically unwell.[1] Although he was not the first to set ventilation standards, Reid criticized existing standards for only defining the minimum required to satisfy metabolic needs, rather than maintaining thermal comfort or an air quality that is perceived as pleasant. These low rates, he wrote, '*would not give the comfort and maintain the constitution in such good condition as a larger allowance*' (Reid 1856, p. 153). To determine the air supply required to achieve an '*agreeable and refreshing atmosphere*' Reid undertook experiments with human participants inside closed chambers in which the air supply could be closely controlled. In these experiments Reid relied largely on the subjective feedback from

[1]1837–38 (277) Ventilation of the House. Letter from Dr. Reid to the Viscount Duncannon, in reply to observations addressed to His Lordship by Sir Frederick Trench.

participants, self-reporting on their experience of indoor climate and atmosphere to which they had been exposed. He asked for feedback on the participants' perceived thermal comfort and air quality '*to ascertain the effect of a given supply of air, at a regulated temperature, renewed in the manner he had proposed*'[2] Reid's finding was that 10 ft^3 per min were an '*ample allowance for an adult*' (Reid 1856, p. 153). During warm weather the rate would need increasing to 40–60 ft^3 per min if a comfortable range of temperatures was to be maintained without the use of artificial cooling,[3] but he also observed that the ventilation rate could only be raised to a certain level, before strong internal air currents, rather than high air temperatures, became the main cause of thermal discomfort (Reid 1856, p. 153).

Inside his private laboratory in Edinburgh Reid constructed various experimental rooms to test different ways of introducing and extracting air in sealed rooms.[4] For Reid sealed rooms with controlled stack ventilation were a means to achieve a tighter control over the climatic and atmospheric conditions than naturally ventilated rooms with operable windows (Reid 1855, p. 208). Air was admitted and discharged exclusively through perforated ceilings, walls and floors. In one of these rooms fresh air was introduced from above through a perforated ceiling, and in another the perforated surface was extended along the walls, allowing the incoming air currents to be distributed over an even larger area (Reid 1855, p. 208). In yet another chamber air was admitted through a perforated floor and discharged via the ceiling. These studies investigated how thermal comfort could be achieved through the different technical arrangements and by optimising the environmental management. Aiming to experimentally determine the conditions at which people felt comfortable, participants were placed inside these rooms and tasked with reporting on their experience of the state of the atmosphere and the physical sensation produced by air currents of varying velocities, degree of diffusion and temperature. These experiments were not dissimilar to the climate chamber studies undertaken by the Willis Carrier in the 1940s (Cooper 1998). Through a system of supply and discharge ducts "*air could be made to enter and be withdrawn in any required proportion*" (Reid 1856, p. 153). These ventilation principles and the working methodology that Reid has used to test them, provided the foundations for the inquiries that Reid had undertaken in Westminster between 1835 and 1852.

[2]Letter from Reid to the Council of the University College, London, 1837 in Testimonials regarding Dr. Reid's qualification as a lecturer on chemistry and teacher of practical chemistry, April 1837 (UCL Library, Hume Tracts).

[3]1837–38 (277) Ventilation of the House. Letter from Dr. Reid to the Viscount Duncannon, in reply to observations addressed to His Lordship by Sir Frederick Trench.

[4]Reid (1855, p. 208); Reid, David, Ground Plan of Dr. Reid's premises, Roxburgh Place, Edinburgh (Parliamentary Archives, OOW/23).

2 The Testing of a Model Debating Chamber

In August 1835 Reid was invited, alongside other scientists and engineers in the field of ventilation and heating, to advise a House of Commons Select Committee appointed to make an inquiry into ways of ventilating the New Palace of Westminster. He outlined the concept for a model debating chamber, building on the findings of his earlier physiological studies. This chamber was completely sealed and the fresh air was admitted entirely through the floor and extracted through the ceiling. It had two large stacks, one serving as a central air inlet, the other as an outlet for used air. A coke fire inside the discharge shaft was to provide the motive power required to extract vitiated air but also to force fresh air from the top of the inlet shaft into the debating chamber (Reid 1844, p. 120). To prevent MPs feet and legs being exposed strong currents, air currents were diffused by covering the entire floor area with small apertures. Reid also submitted a proposal for applying these principles to the Temporary Houses of Commons, which had been erected in Westminster by the architect Robert Smirke in the winter of 1834 after the original Palace had been destroyed in a fire. In their final report the Select Committee did not recommend any specific system to be adopted in the New Palace and Smirke also rejected his plans for testing the system in the Temporary House.[5] However, the Committee approved Reid's proposal to erect a mock-up of the model debating chamber to provide it with '*any additional evidence as to the sound and ventilation might be obtained by actual experiment*'.[6] The test chamber was erected in Spring 1836, next to the laboratory in Edinburgh. The air was admitted through 50,000 holes within the floor and passed into a cavity above the perforated ceiling before entering a duct, which was connected to the discharge shaft of Reid's laboratory.[7] From early summer 1836 Reid used a similar methodology as in his earlier experiments to empirically evaluate the ventilation from physiological perspective. In a letter to the Commissioners of Woods and Forest, dated 28 March 1838, Reid wrote that the chamber was constructed '*with the view of imitating the various circumstances that present themselves during the debates*'.[8] This included studies investigating how the indoor atmosphere and climate was affected by large crowds or by the heat and fumes of gas lighting systems used during debates at night. First experiments, involving groups of over 100 participants, were undertaken in summer 1836. In these studies participants were exposed to different artificial climatic and atmospheric conditions by carefully regulating the

[5]Select Committee on the Ventilation, *Report of the Select Committee on the Ventilation of the Houses of Parliament*, (HC 1835, 583) pp. iii–iv.

[6]1835 Ventilation Committee, Q581–3.

[7]'Philosophical Society', *Caledonian Mercury*, 28 July 1836.

[8]1837–38 (277) Ventilation of the House. Letter from Dr. Reid to the Viscount Duncannon, 28 March 1838.

temperature, relative humidity or adjusting the intensity of internal air currents.[9] The *Caledonian Mercury* wrote that the '*air was completely renewed by a slow and insensible current every five minutes, and the various changes so gradually induced, that it was impossible to tell when they commenced*' and that the chamber was '*filled with warm and cold air, and partially charged with ether and nitrous oxide, at different times*'[10] In other experiments the air supply was switched from the floor to the ceiling and participants were tasked with providing evidence of changes in the thermal sensation produced by these currents.[11] The descending currents reportedly improved the ventilation from a thermal comfort perspective as the feet and legs of participants were no longer directly exposed to currents from below, in particular during periods when the chamber was crowded and the ventilation rate had to be raised in order to maintain a good air quality or prevent overheating.[12] The test chamber was equipped with a system of flues and dampers, which allowed the direction and intensity of the incoming air currents to be regulated.[13] These experiments followed an approach that resembles very closely that used in modern psychophysics. According to Ralph Galbraith Hopkinson, psychophysics as a scientific field aims to gain an understanding of how human beings respond to or perceive their thermal, luminous or acoustic environment. Not dissimilar to Reid, Hopkinson found that this required scientists to '*learn how to use people as meters to register for us their experience in the environment*' (Hopkinson 1963).

3 The Temporary Houses of Commons

In autumn 1836, following the test in Edinburgh and complaints by MPs about Smirke's ventilation system, Reid was finally granted permission to apply his system to the Temporary Houses of Commons.[14] This enabled him, for the first time, to test his principles under real-life conditions, rather than within the artificial settings of his laboratory. From January 1837 until Spring 1851, when the House finally moved into their new permanent debating chamber, Reid's ventilation system was subject of a number of scientific evaluations, which involved experimental

[9]American Dwelling, pp. xv–xxxvii; Reid (1844, pp. 86 and 177; 1856, pp. 164–177); Letter from Dr. Reid to the Viscount Duncannon, 28 March 1838.

[10]'Philosophical Society', *Caledonian Mercury*, 28 July 1836.

[11]Letter from Dr. Reid to the Viscount Duncannon, 28 March 1838; Reid, Illustrations, p. 273.

[12]Letter from Reid to Duncannon, 28 March 1838.

[13]Drawings of the damper system were included in 1837–38 (277) Letter from Dr. Reid to the Viscount Duncannon, 28 March 1838.

[14]Letter from the Treasury to the Commissioners of Woods, 26 August 1836 (National Archives: Work 11/12, Nr. 7); Letter from the Treasury to the Office of Woods, 26 August 1836 (National Archives, work 11/12, nr. 7); Letter from the Office of Woods to Reid, 13 September 1836 (National Archives, work 11/12, nr. 8); 'Miscellanea', *Champion*, 3 October 1836, p. 1; 'London', *Caledonian Mercury*, 17 October 1836, p. 2.

studies, measurements and interviews with regular users, such as MPs and reported.[15] The latter was used as part of investigations into user perception. The ventilation system of the Temporary House cannot be discussed in this paper. An in-depth study of the Temporary Houses of Parliament can be found in *Architectural History* (Schoenefeldt 2014). Similar to Reid's original proposal and the mock-debating chamber, the entire floor, including that in the galleries, was perforated, allowing fresh air to be admitted over the largest possible area (see drawing). Through this, Reid argued, the fresh air currents were diffused and would only become uncomfortably strong when the chamber was exceptionally crowded and the ventilation had to be boosted to prevent overheating or to maintain the required supply of fresh air.[16] The quantity of air passing through the chamber regulated by adjusting the quantity of hot air exhausted through a large air shaft (Fig. 1).

In addition to scientific evaluations, the attendants of the ventilation constantly monitored the internal climatic conditions and collected oral feedback from MPs regarding thermal comfort and air quality. During every debate temperatures, the number of people inside the chamber and the position of the valves used to regulate the ventilation rate were recorded hourly in log-books. The operation of the ventilation was supervised by the Sergeant-at-Arms, Sir William Gosset, who directed the chief attendant of the ventilation, Benjamin Riches, to make adjustments.[17] Gosset dealt with complaints from Members and passed orders to Riches (Reid 1844, p. 325). Thermometers were located, among others, inside the galleries on the east and west side and behind the Speaker's chair on the main floor below.[18] The Department of Woods and Forests transmitted specimens of the collected data to Reid in Edinburgh on a weekly basis, to maintain a constant check on the state of the ventilation. The oral feedback from MPs was collected to monitor aspects of the environment that were highly subjective and/or could not be easily measured with the available technology, such as air purity or the physiological sensation of air currents. Apart from temperature and humidity, no measured data was collected as part of the routine operations.[19] In his *Illustrations of the Theory and Practice* Reid highlighted carbon dioxide levels could not be monitored as *easily* as temperature,

---

[15]Note: Some of the experiments are described in: Report from the Select Committee on Lighting the House (HC 1839, 501); Report from the Select Committee on Ventilation of the New Houses of Parliament (HC 1842, 536); Report of the Select Committee on the Houses of Parliament (HC 448, 1844); Arnott, Memorandum of an inspection made by me, this Day, the 22nd of December 1837, of the arrangement for ventilating and warming the House of Commons' in Hansard HC deb 23 December 1837, vol. 41 cols 329–332.

[16]Illustrations, 1844, p. 274; 'House of Commons', *Sheffield Independent,* 12 November, 1836, p. 1; 1837 (21) Ventilation of the House, Letter from Dr. Reid to Lord Duncannon, 4 February 1837, relative to the acoustic and ventilating arrangements lately made in the House of Commons.

[17]1839 Select Committee, Q744.

[18]Reid (1844, p. 327); 1839 Lighting Committee, pp. 75f.

[19]Select Committee on Lighting the House, Report from the Select Committee on Lighting the House (HC 1839, 501), Q622.

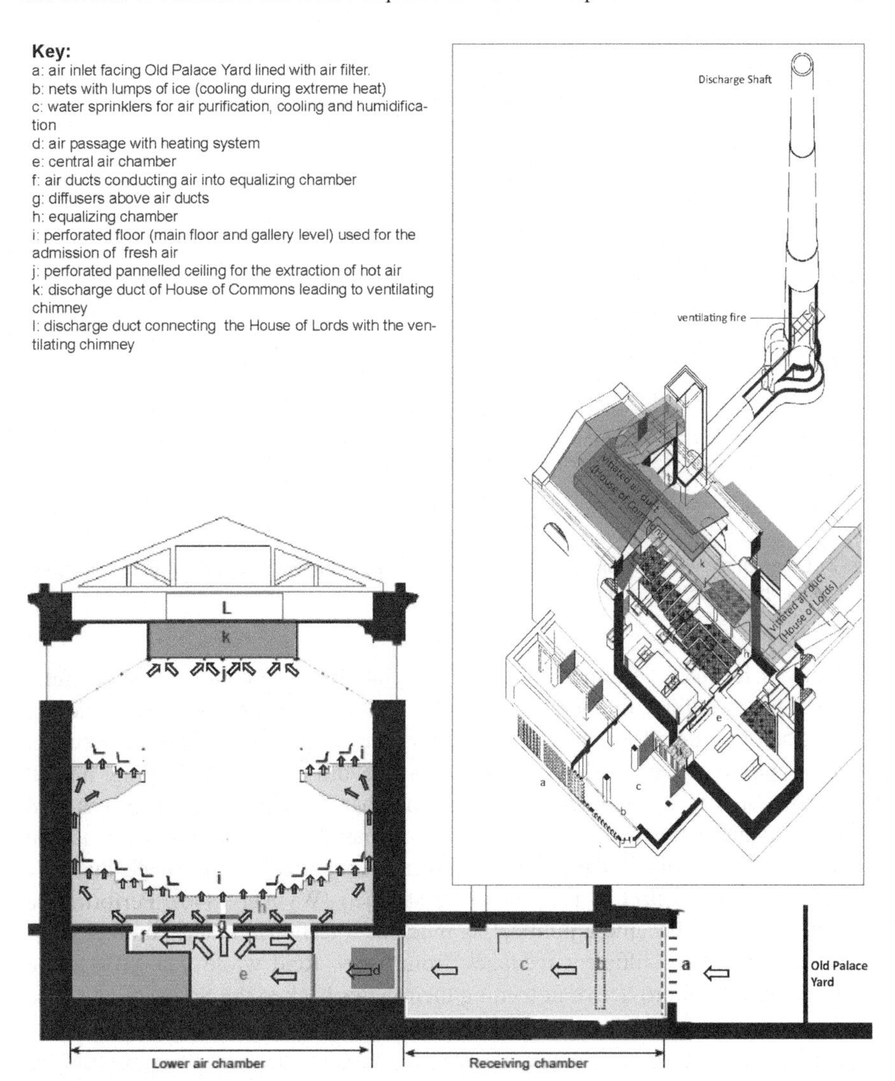

Fig. 1 Cross-section and axonometric projection, showing the stack ventilation system of the Temporary House of Commons

as it required chemical analysis of air samples (Reid 1844, pp. 65–69). The consultation of MPs within the chamber became an integral part of the environmental management regime, complementing the measured data (Wyman 1846) (Fig. 2). The Earl of Shelburne referred to it as a '*system of complaint*'.[20] In addition to the routine management procedures, ad hoc changes were made based on evaluations

[20]Report of the Select Committee on the Houses of Parliament (HC SC 448, 1844).

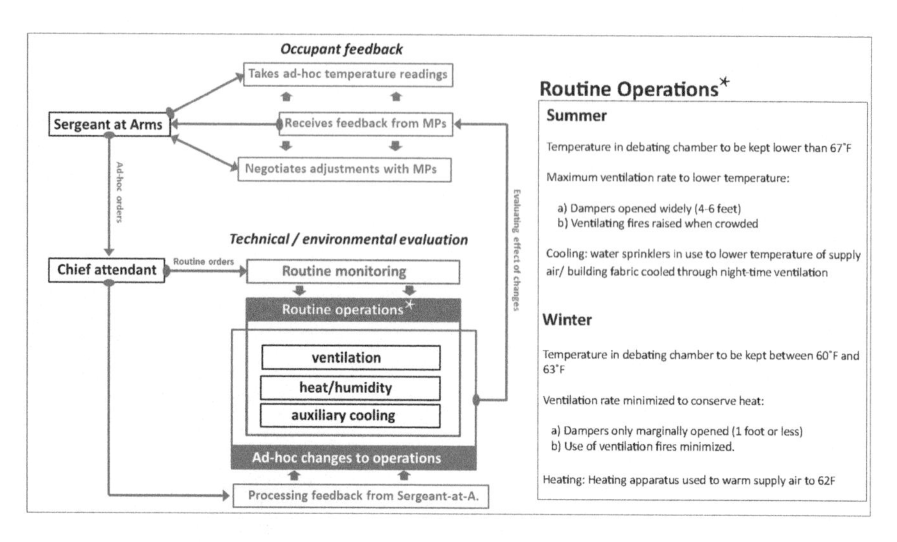

Fig. 2 Diagram showing the operational regime, including occupant feedback system, adopted in the Temporary House of Commons

of occupant feedback. The perceived air purity and thermal comfort was carefully studied. During sittings the Sergeant-at-Arms consulted MPs on their perception of comfort before orders for adjustments were passed. In practice the use of occupant feedback posed some challenges. One was the difficulty of achieving mutual agreements among all Members within the chamber. Gosset noted that it was hard to make ad hoc adjustments to the ventilation based on a consensus of thermal comfort and air purity.[21] Up to 100 adjustments reportedly were made during a single sitting in an effort to keep the MPs satisfied (Wyman 1846). Performance evaluation thereby became a political as much as a technical affair. Another challenge was obtaining sufficient feedback from users. Reid wrote that MPs at the beginning had been pro-active in feeding their views back to the Sergeant-at-Arms, but over time their participation declined significantly.[22]

4 Technical Refinement Based on User-Perception

This monitoring regime had not existed from the start but it had evolved gradually over five years. In an early technical report on his system from February 1837 Reid highlights that he originally avoided the inclusion of direct user-feedback in the monitoring regime. Observations inside the mock debating chamber revealed that it

[21]1839 Lighting Committee, Q758.

[22]Select Committee on the Houses of Parliament, Report of the Select Committee on the Houses of Parliament (HC 448, 1844) Q324–27, 387.

was too difficult to achieve mutual agreements on thermal comfort among large groups. He believed that it would be sufficient to set an average temperature and regulate the ventilation rate according to the number of MPs attending.[23] A manual with series of set temperatures was provided (Fig. 2), but feedback from Members, showed that the perceived temperature was strongly affected by the velocity of the air rising through the perforated floor. When the House was crowded, for instance, the freshness of the atmosphere could only be maintained if the ventilation rate was raised to a level where it produced undesirably strong currents. Under these circumstances the temperature of the supply air needed to be raised to 70 °F before Members ceased complaining about the sensation of chilly currents.[24] Observations inside the Temporary House reconfirmed the thermal comfort implications of a system that relying entirely on perforated floors for the air supply. In his original plans Reid envisaged to address issues with currents, by allowing to supply to be switch from the floor to the ceiling. Apart from a brief trial, the stack-driven system in Westminster was constantly operated in an upward mode, which was largely due to the gas-light below the ceiling. The descending fresh air current was found to carry heat and fumes from these lights into the House.[25] In a mock debating chamber, on the contrary, Reid had adopted an arrangements where the gas flames were isolated from the atmosphere of the chamber through a glass ceiling, but he was not permitted to implement the same arrangement at Westminster. Trials with different lighting arrangements were undertaken, but the issue was never successfully resolved.[26]

In early 1837 only the air temperature was measured, but after a few months it became evident that Reid had been naïve in his assumption and that a more complex monitoring regime and which involved reviewing feedback from MPs was required to improver user satisfaction. Dissatisfaction with the internal conditions drove various MPs to criticize Reid's system and drove subsequent efforts to refine the system and its management.[27] As temperature measurements on their own were also not sufficient to explain the environmental conditions that the MPs were actually experiencing, the scope of measurements was gradually increased. In 1838, following complaints about an excessively dry and dusty atmosphere, Reid tasked his deputy in London with keeping a register of the relative humidity, which had

[23]Letter from Reid to Duncannon, 4 February 1837; D. Reid, *Brief outlines illustrative of the alterations in the House of Commons, in reference to the acoustic and ventilating arrangements*, Edinburgh, 1837.

[24]Report of the Select Committee on the Houses of Parliament (HC 448, 1844), Q557.

[25]Letter F. Trench to Lord Duncannon, 6 May 1838, in Ventilation and Lighting of the House, 1837–8, 358, p. 5.

[26]Select Committee on Ventilation and Lighting of the House, Second report from the Select Committee on Ventilation and Lighting of the House (HC 1852, 402); Select Committee on Lighting the House, Report from the Select Committee on Lighting the House (HC 1839, 501).

[27]'Ventilation of the House of Commons', *Morning Post*, 29 December 1837, p. 1; Savage, James, 'To the editor of the Times', *The Times*, 21 March 1837, p. 5; 'To John Bull', *John Bull*, 8 September 1839, p. 429; Trench, Frederick, 'The new light' *Morning Post*, 19 August 1839.

previously not been monitored.[28] Admitting that a more advanced management regime was required,[29] Reid in 1843 introduced additional thermometers above the ceiling and within the fresh air inlet. This additional data was to provide a better understanding of the effect of weather changes on the system's performance. New columns were added to the original log-sheets for additional data, covering air velocity, use of water and ice for cooling, and fuel for the heating and ventilating furnace. This data was to "*enable the attendants to acquire experience in the various contingencies which they have to meet, and particularly, to enable them to anticipate, as far as possible, every expected change of atmosphere*" (Reid 1844, p. 325). This provided the blueprint for the log-book used for the monitoring of the Permanent House of Commons from 1853 through to 1928.[30] In addition to a more comprehensive monitoring system, complaints also drove Reid to gradually introduce a system of full-climatic control. When the ventilation went operational in winter 1836 it was only equipped with a heating apparatus, but it evolved into a more sophisticated environmental system, which included arrangements for air filtration, regulating atmospheric humidity, and the cooling of the supply air (Fig. 3).[31]

5 The Temporary House of Lords as the Test-Bed for an Alternative Approach, 1838–47

Two years after Reid had applied his system in the House of Commons, he was also commissioned to make alterations to the ventilation in the Temporary House of Lords. Numerous complaints about the poor air quality and thermal discomfort had been made since December 1834 and in summer 1835 became the subject of a parliamentary inquiry.[32] In 1839 Reid was commissioned to address the issue (Reid 1844, p. 271), but he also used the House of Lords as the test-bed for a new approach to environmental management that was different from the House of Commons. He stated that continued disagreement among individual MPs about thermal comfort made it impossible to achieve a high satisfaction rate if the climate within the chamber was uniform throughout. In the House of Lords his objective was to explore how far levels of satisfaction could be increased by delivering local climates within different parts of the chamber.[33] Although the climate could not be

[28]Letter from Trench to Duncannon, 22 April 1838.

[29]Letter from Reid to Ducannon, 28 March 1838.

[30]E.g. Registers of temperature control and ventilation for the House of Commons 1853–1928 (Parliamentary Archives, OOW/5).

[31]Letter from Reid to the Chancellor of the Exchequer, 17 June 1837; Reid, Brief Outlines, p. 14; Reports from the Select Committee of the House of Lords appointed to inquire into the progress of the building of the Houses of Parliament (HL 1846, 719).

[32]'The House of Parliament', *Morning Post*, 26 December 1834, p. 4.

[33]Select committee on Houses of Parliament, Report of Select Committee on Houses of Parliament (HC 1844, 448), Q317–23.

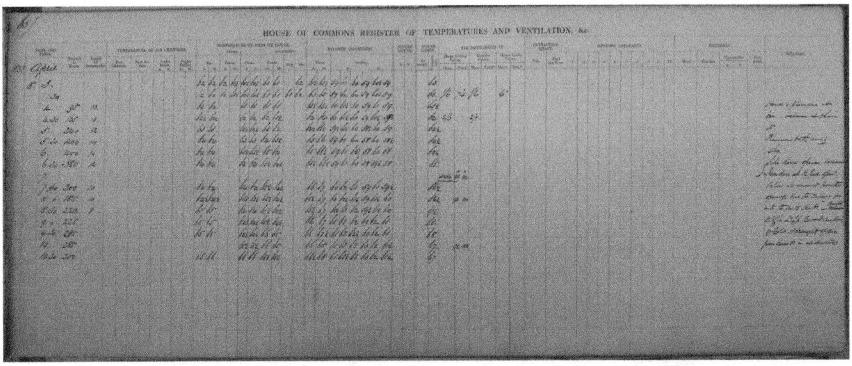

Fig. 3 Page (8 April 1853) from the original log-book used to record measured data and oral feedback from MPs in the Permanent House of Commons (parliamentary archives, OOW/5)

tailored to each individual, the management was able to create some climatic variation throughout the chamber. In the more crowded areas for instance, where there was a greater tendency to overheating, cooler air was introduced, while areas that were more sparsely populated could be supplied with warmer air. Air of different temperatures and velocities could be introduced at the benches on opposite sides as well as around the bar and the throne, producing cooler or warmer zones within the chamber (Fig. 4). At times the difference in temperature in the air introduced in one section could be as low as 52 °F and as high as 75 °F in another.[34] Fresh air was supplied through a cavity behind the wall panelling, from under the benches and tables and through grills in the risers of the raked floor. Fresh air was purified, humidified, heated or cooled in the tempering chamber below the main floor (Reid 1844, pp. 284–293), and to gain more control over the supply and independently from the pressure of the stack, Reid introduced a separate plenum system.[35] Used air was extracted through a series of circular openings inside the ceiling, which were connected to the discharge shaft of the House of Commons via a large duct (Brayley and Britton 1936, p. 463).

This new strategy did not succeed in increasing user-satisfaction. The idea of managing the different climatic zones based on oral feedback was found to be difficult to implement. The Peers had various debates about the indoor climate, which illustrates how they had perceived the internal conditions. Lord Campbell wrote that the system was successful in maintaining a good air quality but the main problem was insufficient control over the temperature and currents.[36] During a debate in February 1843 he complained that the '*alternate heat and cold of the*

[34]1844 Committee on Houses, Q322.

[35]Select Committee on Ventilation and Lighting of the House, Second report from the Select Committee on Ventilation and Lighting of the House (HC 1852, 402), Q511–542.

[36]Letter from Lord Campbell, 11 September 1843, in Extracts from official documents, reports (UCL Library, Hume Tracts).

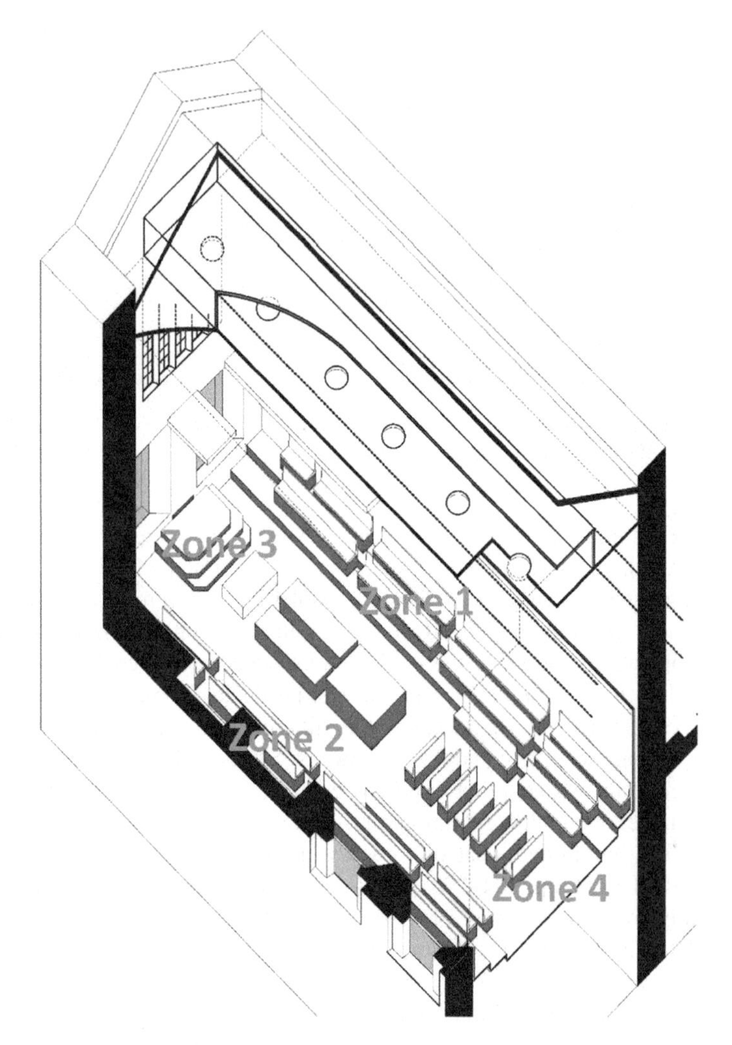

Fig. 4 Axonometric projection showing interior of Temporary Houses of Lords

place made it at one time a cold bath, and at another a vapour bath'.[37] On 5 June 1845 Lord Brougham complained about the wretched state of the atmosphere and Campbell reported that some Peers '*suffered so severely last night from the imperfect ventilation, and the sudden draughts of hot and cold air*'.[38] In 24 April 1846 Lord Brougham notes that the '*Lords were sometimes broiling and sometimes freezing*'.[39] For Reid the main issue was not technical but the communication

[37]The New Houses of Parliament, HL Deb 21 February 1843, vol 66 cc1033–6.

[38]The New Houses of parliament, HL Deb 05 June 1845. vol 81 cc120–2.

[39]Progress of the New Houses of Parliament, HL Deb 24 April 1846 vol 85 cc970–6.

between the Peers and the attendants of the ventilation. During an interview with the 1844 Select Committee Reid stated that the technology permitted a high level of control over the temperature and velocity of air currents, but the attendants relied on regular feedback from individuals occupying different areas to effectively respond to their specific needs. Critical feedback had been very limited, even when the Peers felt uncomfortable in the part they were occupying.[40]

6 Towards the Personalization of Environmental Control

Despite widespread disapproval Reid pursued developing the idea of more personalized climates in the 1840s, whilst working on the ventilation for the Permanent Houses of Commons and Lords in the New Palace of Westminster. Reid started developing a stack ventilation system to be incorporated in Charles Barry's architectural scheme for the New Palace of Westminster in 1840.[41] He developed first concepts for a system providing personalized climates in his early proposals for ventilating the two permanent debating chamber inside the Palace, which were finally implemented in the Permanent Houses of Commons between 1846 and 1852 (Fig. 5). These final arrangements cannot be discussed in detail in this paper, but a more in-depth study has been published in *Studies in the History of Construction* (Schoenefeldt 2016a, 2018). Reid realized another system following similar principles in St. George's Hall in Liverpool.[42] A more complex air handling strategy, following similar principles tested in the mock-debating, was adopted to provide attendance with the flexibility to operate the ventilation in different modes. To allow the system be operated in an upward or downward mode two separate air supplies with their own tempering chambers were introduced, one above the ceiling and another below the main floor. Facilities for addressing individual differences in perceived thermal comfort through the creation of local climates was increased by allowing the climate of each bench to be individually adjusted in response to requests from users.[43]

In contrast to the Temporary House of Commons, where fresh air was admitted uniformly across the entire floor, the floor inlets in the Permanent House were confined, as far as possible, to areas where Members were not exposed to incoming currents. The main air supply was through central floor between the table and bar,[44] but fresh air was also admitted through the perforated floor along the back of the

[40]1844 Select Committee on Houses, Q317–28, 387; Reid, Illustrations pp. 292–293.

[41]A detailed study on the design development process, including the collaboration between Reid and Barry's teams can be found in Schoenefeldt (2016, pp. 175–199).

[42]Mackenzie (1863, pp. 194–208); Reid, Diagrams of the Ventilation of St. George's Hall, To the right Worshipful the Mayor and Corporation of Liverpool and the Law Courts Committee, under whose direction and superintendence St. George's Hall and the New Assize Courts have been constructed, 21 May 1855 (Liverpool Record Office).

[43]1844 Committee on Houses, Q317–28, 387.

[44]Interview with Reid, 30 April 1852, in 1852 Select Committee (SC 1852 Q3545).

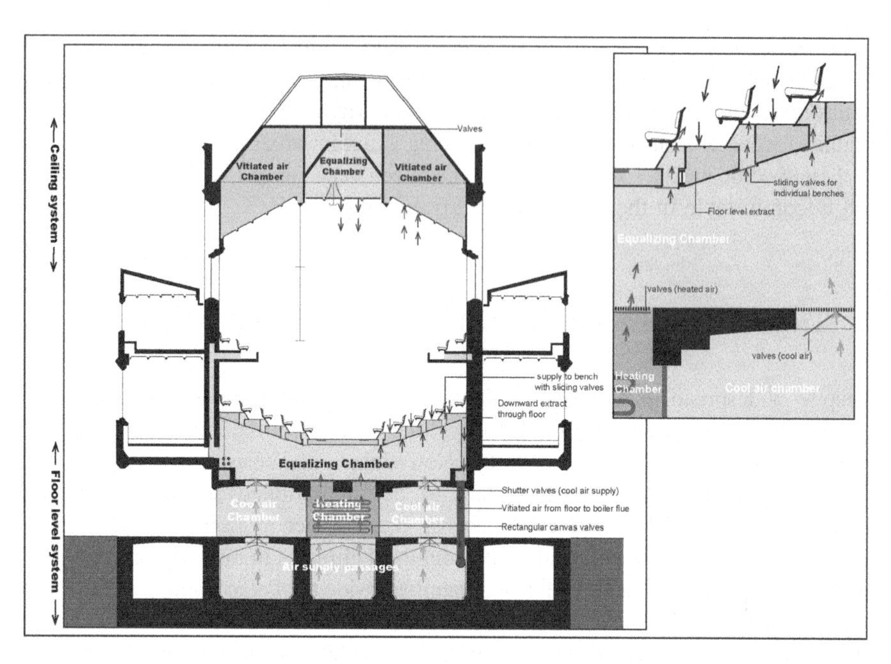

Fig. 5 Cross-sections of House of Commons, showing how air was handled at ceiling and floor level

benches and through the risers in the gangways. Separate inlets were provided for the Speaker's chair and within the crowded areas around the table.[45] The supply in each of these areas was designed to be regulated independently, involving over 60 individual sliding valves that could be manually adjusted by attendants from inside the equalizing chamber. Working drawings and written notes from 1851 illustrate that each bench and riser was provided with a separate supply duct and valve.[46] The humidity, temperature and velocity of the fresh air admitted was regulated inside the Equalizing Chamber, below the main floor.[47] Thermometers and hygrometers were situated inside the equalising chamber, allowing the attendant to closely monitor the temperature and relative humidity of the air. The temperature was adjusted by regulating the relative quantity of cool and heat air admitted through separate valves from the air tempering chamber on the level below. This chamber had central heating compartment with hot water pipes, surround by a cool air

[45]Half-plan, showing arrangements of air valves in the ceiling of the basement and the floor of the equalizing chamber, dated 5 April 1847 (Work 29/3026); Section through chamber, 5 April 1847 (Work 29/3029).

[46]Plan showing air supply tubes serving individual benches, April 1847 (National Archives, Work 29/3046) Ceiling of equalizing chamber under house with supply tubes, valves and flaps, 23 June 1851, (National Archives, Work 29/3100).

[47]Second report from the Select Committee on Ventilation and Lighting of the House (HC SC 1852, 402).

chamber, through which unheated air could pass directly from the basement into the equalizing chamber. The fresh was supplied primarily from the top of the Clock Tower and was conveyed to the House through air passages in the basement.

7 An Unsuccessful Experiment

Transcripts of debates and parliamentary papers reveal that the MPs were highly dissatisfied with the climatic conditions inside the new House, and as before, were very vocal demanding remedial measures.[48] Between 1852 and 1854 the new ventilation became the subject of inquiries by four separate parliamentary committees and several independent scientific studies were commissioned, investigating the causes of discomfort.[49] These inquiries revealed that this sophisticated management strategy that Reid had proposed was never been fully put into practice. This was partly due to the (1) design of the gas lighting, which, as in the Temporary House, prevented the use of the ceiling supply,[50] and partly as (2) critical features never completed and due to (3) complexity of the monitoring and management regime. Managing a complex operations which were required to respond to changes in the weather or number of MPs, let alone to feedback from users, was a major managerial task. As these operations were done completely manually, without the aid of modern computerized monitoring and control, the system's performance relied heavily on the effective co-ordination of the recording and adjustment procedures. A large volume of data had to be gathered, interpreted and communicated alongside adjustments to ventilation dampers and the heating, cooling and air filtration systems. Failure to satisfy the MPs, however, led to the system being abandoned in 1854, only two after it went first operational (Schoenefeldt 2018a, 2019). It was the demands of the occupant, in the case an occupant of unusually high influence and power that led to the fall of Reid's system.

[48]E.g. Imperial Parliament, Daily News, 5 February 1852, p. 3, Memorandum submitted by Reid to the Commissioners of Works, 7 February 1852 (work 11/14 nr. 678); Ventilation of the House, HC Deb 06 February 1852 vol 119 cc231–4 231, Ventilation of the House, HC Deb 11 February 1852, vol 119 cc400–16.

[49]Second report of Mr. Goldsworthy Gurney on the ventilation of the new House of Commons. (HC 1852, 252 (371)); Second report from the Select Committee on Ventilation and Lighting of the House. HC 1852 (402); First Report on the State of the Warming of the Warming, Ventilating, and Lighting arrangements throughout the building, Meeson, December 1852 (works 11/14, nr. 768–81); First report from the Select Committee on the Ventilation of the House (HC 1854, 149), Report from the Select Committee of the House of Lords, appointed to inquire into the possibility of improving the ventilation and the lighting of the House (HL SC 1854, 384).

[50]Memoranda submitted to chief commissioner of the office of Works, 7 February 1851 (work 11/14 nr. 678); Alfred Meeson, First Report on the State of the Warming, Ventilating, and Lighting arrangements throughout the building, December 1852. (work 11/14 nr. 768–81).

8 Conclusion

This paper has explored how user-perception has acted as a major driving force in the development of Reid's design for the ventilation of Houses of Commons and in evaluating its performance. User perception became a central measure in evaluating the effectiveness of ventilation system from the point of thermal comfort or air quality. Reid tested environmental monitoring regimes that combined measurements with the review of oral feedback from users. This was adopted first to account for various aspects of the environment that had not been monitored but also to gain an understanding of the perceived environment and how this differed from what the measurements were suggesting. In *The Architecture of the Well-Tempered Environment* Raynar Banham stressed the reliance of 19th century scientists on the human senses in their efforts to study of the environment of enclosed spaces (Banham 1984). The 'measurable' and the 'perceived', however, were integrated not only to gain a comprehensive understanding of environmental conditions inside the chamber, but also to monitor these climate and atmosphere affected occupants. Feedback from MPs was used to track changes of thermal perception over time, assess general levels of satisfaction among MPs present, and to identify needs of individuals. Climate control became a highly political process, involving attempts to manage the shared climate according to the feelings of the majority on one side, and to accommodate the demands of individuals through local climates on the other. This included experiments with 'consensus-based' and 'personalised' approaches to environmental control, which have remained the subject of conflicting philosophies of thermal comfort up to the present day. The BRE's Environmental Building in the UK, for instance, combines a central BMS system with remote control. The latter permitted users to adapt the indoor environment to their personal preferences by manually overriding the central system. The work in Westminster reflected a highly developed understanding of the role of the occupant-environment relationship in environmental design, which resembles very closely that of current methods post-occupancy evaluations, in which user-surveys are combined with the environmental monitoring. Reid's work, however, predated the modern methods of post-occupancy evaluation developed in the 1960s (Preiser and Hardy 2015), including those used in Ellie Morgan's Wallasey School, by 130 years.

References

Banham R (1984) Architecture of the well-tempered environment, 2nd edn. University of Chicago Press, Chicago

Brayley E, Britton J (1936) The history of the ancient palace and late houses of parliament at Westminster. John Weale, London

Cooper G (1998) Air-conditioning America: engineers and the controlled environment—1900–1960. Baltimore

Hopkinson RG (1963) Architectural physics: lighting. HMSO, London
Mackenzie W (1863) On the mechanical ventilation and warming of St. George's Hall, Liverpool. In: Proceedings of the institution of mechanical engineers, pp 194–208
Preiser W, Hardy A (2015) Historic review of building performance evaluation, in architecture beyond criticism. Routledge, London
Reid D (1844) Illustrations of the theory and practice of ventilation, London
Reid D (1855) The revision of architecture in connection with the useful arts. Builder, 5 May 1855, pp 13–639
Reid D (1856) Eight lectures by David Boswell Reid on 'Progress of architecture in relation to ventilation, warming, lighting, fire-proofing, acoustics, and the general preservation of health'. Smithsonian annual reports, pp 147–186
Schoenefeldt H (2014) The temporary houses of parliament and David Boswell Reid's architecture of experimentation. Arch Hist 57:175–215
Schoenefeldt H (2016) Architectural and scientific principles in the design of the palace of Westminster. In: Gothic revival worldwide A. W. N. Pugin's global influence. University of Leuven Press
Schoenefeldt H (2016a) The lost (first) chamber of the house of commons. AA Files 72:44–56
Schoenefeldt H (2018) The historic ventilation system of the house of commons, 1840–52: revisiting David Boswell Reid's environmental legacy. J Soc Antiq 98:245–295. https://doi.org/10.1017/S0003581518000549
Schoenefeldt H (2018a) Powers of politics, scientific measurement, and perception: evaluating the performance of the houses of commons' first environmental system, 1852–4'. In: Joyce H, Gillin E (eds) Experiencing architecture in the nineteenth-century, London: Bloomsbury, pp. 115–129
Schoenefeldt H (2019) The house of commons: a precedent for post-occupancy evaluation. Build Res Inf 47(6): 635-665. https://doi.org/10.1080/09613218.2019.1547547
Wyman M (1846) A practical treatise on ventilation. Chapman Brothers, London

Printed in the United States
By Bookmasters